Evolving Genres in Web-mediated Communication

Linguistic Insights

Studies in Language and Communication

Edited by Maurizio Gotti,
University of Bergamo

Volume 140

PETER LANG
Bern · Berlin · Bruxelles · Frankfurt am Main · New York · Oxford · Wien

Sandra Campagna, Giuliana Garzone,
Cornelia Ilie & Elizabeth Rowley-Jolivet (eds)

Evolving Genres in Web-mediated Communication

PETER LANG
Bern · Berlin · Bruxelles · Frankfurt am Main · New York · Oxford · Wien

Bibliographic information published by die Deutsche Nationalbibliothek
Die Deutsche Nationalbibliothek lists this publication in the Deutsche Nationalbibliografie; detailed bibliographic data is available on the Internet at ‹http://dnb.d-nb.de›.

British Library Cataloguing-in-Publication Data:
A catalogue record for this book is available from The British Library, Great Britain

Library of Congress Cataloging-in-Publication Data

Evolving Genres in Web-mediated Communication / Sandra Campagna, Giuliana Garzone, Cornelia Ilie & Elizabeth Rowley-Jolivet (eds).
pages cm. – (Linguistic insights : Studies in language and communication, ISSN 1424-8689 ; Volume 140)
Includes bibliographical references.
ISBN 978-3-0343-1013-0
1. Discourse analysis–Social aspects. 2. Technological innovations–Social aspects. 3. Web 2.0–Social aspects. 4. Social media. I. Campagna, Sandra.
P302.84.E96 2012
401.41–dc23

2012038395

Published with the support of the Turin Chamber of Commerce and of the Dipartimento di Scienze della Mediazione Linguistica e di Studi Interculturali, Università degli Studi di Milano, and based on Seminar 66, "From downloading to uploading: new spaces and new voices in web authoring", and on Seminar 68, "Evolving Genres and Genre Theory: Focus on Specialized Communication in English across Contexts and Media", held at the Tenth European Society for the Study of English (ESSE) Conference, Turin 24-28 August 2010.

ISSN 1424-8689 pb.
ISBN 978-3-0343-1013-0 pb.

ISSN 2235-6371 eBook
ISBN 978-3-0351-0436-3 eBook

Hochfeldstrasse 32, CH-3012 Bern, Switzerland
info@peterlang.com, www.peterlang.com, www.peterlang.net

Contents

Web 2.0

Exploring the Blogosphere

SANDRA CAMPAGNA / GIULIANA GARZONE / CORNELIA ILIE /
ELIZABETH ROWLEY-JOLIVET

Introduction

This volume deals with discourse genres in Web-mediated communication, focusing in particular on genre change and evolution under the pressure of technological renewal and the availability of new affordances. Several of the chapters are based on papers presented at the 10th Conference of the European Society for the Study of English, in Seminar 66 'From Downloading to Uploading: New Spaces and New Voices in Web Authoring', convened by Sandra Campagna and Elizabeth Rowley-Jolivet. The remaining chapters are based on a selection of the papers presented at the same Conference in Seminar 68, 'Evolving Genres and Genre Theory: Focus on Specialized Communication in English across Contexts and Media', convened by Giuliana Garzone and Cornelia Ilie, chosen for their relevance to the main theme of the volume. All the contributions selected have been submitted to blind peer review.

1. Genres in Web-mediated communication: from Web 1.0 to Web 2.0

The last few years have seen a surge in scholarly interest in Web-mediated communication, which has by now overtaken all other media in terms of pervasiveness and volume of information exchanged, and, thanks to ever smarter technological developments, has been gradually incorporating most of such media (just think of the possibility to access video, TV, films and music through the computer).

As tacitly accepted genre norms are gradually being challenged and transgressed due to changing circumstances (such as increased competitiveness), new genre conventions are constantly emerging at the interface between initially different, but often converging, genres. It is symptomatic that earlier homogeneous genres tend to turn into heterogeneous genre hybrids. Newly developed web genres share a number of common features reflecting current expectations of the members of virtual communities of practice: convergent communicative goals meant to establish and uphold interpersonal and institutional relations, enhanced inter-disciplinary and cross-cultural use of participatory affordances, multi-level strategies facilitating open access to and evaluation of information, widespread commitment to encouraging collective linguistic creativity and rhetorical emancipation.

These profound changes and the issues they raise have prompted a surge in scholarly interest, also leading to the inception of new lines of research. The early studies focusing on communication through the Internet were essentially aimed at giving theoretical accommodation to the unprecedented peculiarities of web genres and at adapting existing analytical tools to make them applicable to the new objects of investigation. In particular, there were two aspects that commanded attention: the hypertextual character of Web-mediated communication and the multimodal options made available by computer affordances.

Hypertextuality – today realized as hypermediality – is the constitutive property of Web text, which typically is, in Landow's original definition, "text composed of blocks of words (or images) linked electronically by multiple paths, chains, or trails in an open-ended, perpetually unfinished textuality described by the terms *link, node, network, web,* and *path*" (Landow 1992: 3).

The hypermedial nature of this environment brings with it a number of peculiar characteristics: first of all, granularity, i.e. the fact that in a hypertext the text is broken down into lexias[1] i.e. discrete units – often coinciding with the amount of text that can be displayed on the screen – mutually connected by means of links; recourse to

1 The word *lexia* was introduced into English by Landow (1992: 34) to translate Roland Barthes' term *lexie* (1970/1974), the meaning of which broadly corresponds to that of Derrida's *morceaux* (1972).

multiple consumption modes, i.e. the traditional linear 'reading-as-such' modality and the non-linear 'navigating mode' (cf. Askehave/Ellerup Nielsen 2004, 2005; Garzone 2007), which combined together generate the typical hyper-reading mode (Sosnoski 1999: 135) in which the Web user continuously shifts from reading to navigating and vice versa; co-articulation, i.e. the fact that the user must actively participate in the realization of any web application, giving it order and shape as s/he constructs his/her own navigation path (cf. e.g. Garzone 1997; 2002: 286-287). In actual fact, as will be seen later, co-articulation is possibly the characteristic of the Web environment that has undergone the most meaningful developments, bringing about new applications and new media totally centred on it.

The other important issue extensively investigated in recent years by researchers working in different disciplinary areas, from linguistics and media studies to Human-Computer Interaction sciences, from perceptual psychology to business communication, is multimodality as an option inherent in Hypermedia Computer-Mediated Environments (HCMEs, cf. Hoffman/Novak 1996). In Stöckl's (2004: 9) definition, multimodality is a complex of "communicative artefacts and processes which combine various sign systems (modes) and whose production and reception calls upon the communicators to semantically and formally interrelate all sign repertoires present" (Stöckl 2004: 9). Multimodality is an option made possible to an unprecedented degree by the HCME thanks to its affordances, which give simultaneous access to various media and offer the opportunity to combine different semiotic resources into a single communicative act.

All these characteristics, together with a practically unlimited extension of the participation framework, as the WWW provides a virtually global audience to anything that is published on it, have an inevitable impact on all Internet genres, whether they be derived from pre-existing genres or native to the computer environment, determining – among other consequences – a radical change in the way textual moves are realized, loosening their structural stability, and deconstructing their integrity.

Of course, the great challenge posed to genre theory by the rise of computer-mediated and Web-mediated communication (cf. Lemke 2002; Askehave/Ellerup Nielsen 2005; Garzone 2007) calls into question the traditional purposive, goal-directed view of genre (cf. Swales 1990; Bhatia 1993). The solution proposed, consisting in the incorporation of medium into this notion (Askehave/Swales 2001), has not been considered adequate unless it also integrates the peculiarities described above, all of which have to be taken into account in order to accommodate the distinctive features of Web-mediated genres in genre theory (cf. Garzone 2007). This work of finetuning existing categories, which since its very beginning has mostly focused on websites and other Web 1.0 applications, is still underway.

In more recent times, the inception and increasing popularity of Web 2.0 applications has added new issues and problems to those already investigated. On all Web 2.0 platforms discursive practices are shaped by direct intervention on the part of the users, turning co-articulation into co-construction. As clearly explained by Friedman (2005), the new collaborative platforms are characterized by the increasingly pervasive tendency among web users to *upload* rather than simply *download* information from the Internet, that is to express opinions, exchange ideas, produce materials and globalize the content directly on the Web.

This process of global cooperation often involves bypassing hierarchical organizations or institutions, thus favouring both an active, authoring role on the part of the user and the emergence of new, less constrained, communicative practices. Indeed this participatory agenda is visible in many classic virtual Web communities, such as Second Life, in popular sites like Facebook, YouTube, MySpace and Twitter, as well as in environments such as blogs, discussion forums or online debates; or in spaces where authors actively contribute to the dissemination of knowledge, like Wikipedia. Recently, even more complex social practices have emerged, on a blurred terrain strongly marked and shaped by the emerging authoring voices which not only contribute to the co-construction of information in these New Media, but also impact outside the Web domain by calling into question the role and operation of mainstream mass media.

This collaborative process has several implications: firstly, web authoring involves multiple identities: the user/viewer/reader can be a critical voice, an opinion maker, a cause supporter, a technical advisor. Moreover, this new authoring role has repercussions on more traditional forms of communication, where the role of 'expert-in-the-field', which was in the past clearly assigned to an authority figure, is now challenged by several 'experts-in-the-field' taking the floor.

Second, the blurring of the traditional distinction between text producer and text receiver (or communicator/audience) raises concerns on the part of the discourse analyst: how is textual coherence constructed in the 'new' texts? How does the overtly dialogic nature of the participatory agenda reverberate on the final text and impinge on 'authorship'? A third implication revolves around the growing hybridity of previously discrete discourse genres and communicative contexts. Online news discourse and advertising, for example, are increasingly taking the shape of discussion platforms thanks to users' critical interventions. Active participation on the part of web users is often strongly solicited to co-construct 'infotainment' events, a trait which confirms the trend noted in Fairclough (1995: 10-11) towards an intensified 'marketization' of the media which, due to increasing commercial pressures and competition, are more and more induced to operate "on a market basis within the 'leisure' industry, and one part of that is greater pressure to entertain even within public affairs output".

This participatory agenda challenges traditional genre theory by posing interesting questions related to the emergence of complex, less controlled discursive practices due to this process of co-construction and multiple authoring. Evidence of such complexity is found in various domains, such as, for example, to name just a few, the context of open-source business platforms, online-publishing and so-called participatory journalism. Among these, especially pervasive is the impact of the latter, as the websites of most TV news channels as well as those of the major daily and weekly papers now include space for readers'/viewers' comments and opinions, and strongly elicit their contributions as citizen journalists, inviting them to upload video footage, photos or opinions on news events. The popularity of this kind of non-hierarchical debate in the form of blogs and comment options raises the question of the potential risks of manipulation and of the implications that this development will have for the concept of 'reader /viewer', and for the traditional broadcast or paper versions which continue to exist alongside the electronic version. What is mainly at issue is how these newly-constructed contexts relate to traditionally-shaped communicative practices (and genres).

2. Structure and contents of the book

The book is organised in three sections: the first devoted to traditional web applications, the second dealing with various types of Web 2.0 application, and the third one narrowing down on blogging and micro-blogging as the applications that are most representative of the properties of the new platforms, which offer virtually unlimited scope for individuals' participation, and at the same time – probably because of this – are amongst the most popular.[2]

The first section, devoted to Web 1.0 applications, opens with a theoretically oriented chapter, which focuses on the website, certainly the first and foremost generic construct that comes to mind when talking about Internet communication, given its consolidated status. While its practical applications are virtually ubiquitous, its nature raises a number of theoretical issues. The first chapter, authored by PAOLA CATENACCIO, discusses some of these theoretical issues in a discourse-theoretical perspective. Its starting point is the consideration that, despite being often taken for granted in studies of Web genres, the website as a construct seems to have been given relatively little consideration as an object of serious investigation. Reviewing recent developments in genre studies both from a genre-theoretical and from a systems science perspective, Catenaccio attempts to provide a definition of the website capable of capturing both its self-contained (albeit open-ended) nature, and its function as a 'container' of multiple independent genres. Her conclusion is that the website should be considered as a rhetorical interface enabling access to multiple interrelated genres which jointly build a network featuring dense intertextual references, and in which recontextualisation processes play a key role.

---

2 Unless otherwise indicated, all websites cited in the volume were last accessed on 23/03/2012.

The other chapters in this section deal with various uses to which websites are put in the contemporary world. Today Web-mediated communication has reached a mature stage, having some decades of history behind it, and is being utilized for a number of different purposes, as well as playing an important role in the public life of many countries. So research has also been turning attention to practical applications, and to a consideration of the value and impact of Web mediation on genre use in all kinds of activities in the contemporary world.

A case in point is represented by the new medical and healthcare digital communication genres. E-health is a field of web-mediated communication at the intersection of medical informatics, public health and business, involving health services and information delivered or enhanced through the Internet and related technologies. Health 2.0 can be envisaged as a participatory healthcare process, whereby patients, physicians, providers, etc. use a specific set of Web tools (blogs, Podcasts, tagging, wikis, etc.) to personalise health care, collaborate, and/or promote health education. It has gradually acquired a wider scope since it builds on a commitment for networked, global thinking, aimed at improving health care locally, regionally, and worldwide (Glanz *et al.* 2002). Within this context, current socio-political phenomena like immigration have a significant impact through cross-cultural and multilingual variations in generic realisations of healthcare/medical (e)brochures, whose varying aspects related to ethnicity, race, gender and culture-specific health issues raise serious concerns.

This emerges clearly in ALESSANDRA VICENTINI's chapter, which focuses on the use of websites by the public health service for the purpose of communicating with foreign immigrants. On the basis of a corpus of institutional healthcare websites collected from the site of Italy's Ministry of Health, it aims to find evidence of the recent immigrants' presence in the channels used by Italian institutions to communicate with the general public, with specific focus on the typology of health information. Special attention is drawn to the description and definition of institutional healthcare (e)brochures for foreigners (i.e. the new immigrants to Italy) from a genre-analysis and critical discourse analysis perspective. The investigation reveals a general tendency toward simplification in text structure and rhetorical strategies as well as a heavy shift from the written to the visual semiotic code. Moreover, the reduced content suggests that these (e)brochures are used as a first contact tool to offer new immigrants help in accessing public health-care facilities, rather than as an institutional document providing answers or describing diseases and conditions, as was the case with the previous versions of the same brochures.

The next chapter concentrates on the political use of web-mediated communication, whose potential has been clear right from the very early stages of the spread of ICTs (cf. e.g. Archibald 2007): Governments make use of information and communication technologies (ICTs) to improve public sector organisations' activities and provide better public services, but also to stimulate citizens' participatory role, rethinking organisational structures and processes, and changing behaviour so that public services are delivered more efficiently. A wide range of administrative, information-oriented, as well as self-promotion-oriented genres, are normally displayed on government websites. The design and layout of officially and openly displayed genres reflects the main focus and scope of the services provided through e-Government. A relevant case study is that discussed in her chapter by BETTINA MOTTURA, who deals with the use of the Internet by the Government of the People‘s Republic of China. Its website (online since 2006) is identified as one of the crucial media to perform two public administrative functions: spreading information to the general public and collecting feedback from society. Therefore, the analysis reveals the choice of linguistic resources recently adopted by the Chinese public administration in order to perform this two-way communication flow with the citizen. Findings indicate the simultaneous presence of traditional bureaucratic genres and of textual forms derived from other professional communities, identifying the functions they serve. In particular, Mottura looks at the genre of the online interview, devoting specific attention to three interviews with Prime Minister Wen Jiabao conducted in recent years which are examined with the underlying hypothesis that their linguistic features are heavily influenced by the distribution channel.

In the last chapter of the section devoted to Web 1.0 applications, the focus shifts to NGOs' websites, with special regard for rhetorical strategies and their realization in the hypertextual environment. In her chapter CHIARA DEGANO investigates variation in argumentative discourse as a consequence of genre migration to the Web, using the case of NGOs' campaigns as a case study. The analysis addresses issues connected with lack of linearity in hypertexts, and in particular its impact on textual coherence, and the related loss of the writer's control on the *disposition* of rhetorical moves. It concludes that coherence relations – and the effectiveness of argumentation with them – are indeed put at risk by the potentially unstructured character of the hypertext. Especially problematic is the information overload, combined as it often is with the failure to create recognizable textual patterns across the levels of the hypertext, a problem that, however, can be avoided by means of a 'dynamic coherence planning' of the hypertext. In the cases analysed, dynamic coherence is achieved globally by suggesting a pattern of reading which reflects a hierarchical organization of issues and by having recourse to an effective distribution of argumentative moves across levels.

In the following section the focus shifts to more innovative Web 2.0 applications, tackling issues that have only recently risen to researchers' attention, given that these genres are relatively new and have been little studied so far. So, especially in the first two chapters, the focus tends to be on differences with traditional media and on issues connected with the need to adapt existing analytical tools to the new objects of investigation.

The first area discussed in this Section is that of e-Science, in which newly emerging genres are used to promote, in highly distributed network environments, the development of new web-based tools and infrastructures aimed at supporting scientific discovery. Science 2.0 has marked a paradigm shift from the earlier tradition of publishing final results and concluded investigations by well-defined collaborative groups towards a more flexible and open-ended approach, which encourages the development of new digital scientific genres that include the public sharing of raw data, preliminary experimental results, and conjectural information. This is very effectively illustrated in ELIZABETH ROWLEY-JOLIVET's chapter which presents a study of Open Science, a web-based development that has emerged over the last ten years, whose central tenet is the effort to make scientific methodologies, tools, primary data and results freely available on the Internet. These Web-mediated data are mutable, evolving through edits made by a worldwide collaborative community of scientific contributors. Within this context, Rowley-Jolivet focuses on one long-standing scientific genre, laboratory protocols, and compares a selection of them from the collaborative OpenWetWare website with traditional print protocols, looking at the transformations this genre has undergone in its migration from the traditional print to the web-mediated version. This gives her the opportunity to tackle the question of whether traditional genre theory can handle these novel Web-mediated genres, or whether a 're-purposing' of genre theory (Askehave/Swales 2001) is necessary.

This investigation of collaborative, mutable Web-mediated genres is pursued in the following chapter by MARISTELLA GATTO, who looks at Wikipedia, the prototypical Web 2.0 genre, and uses it as a case study to explore the impact of technological affordances on the action of centripetal and centrifugal forces in discourse. In particular, she examines some key questions relating to Web 2.0 as a new writing space highlighting such fundamental notions as *authorship* vs *audience*, *consumers/producers* vs *prosumers*, *upload* vs *download*, *taxonomies* vs *folksonomies*, all considered critical to Wikipedia as a co-operative writing enterprise. In the final part of the study, the evolution of genre in the specific case of the encyclopaedia is investigated by comparing/contrasting different stages in the history of a sample of Wikipedia entries, providing evidence of the action of the centrifugal/centripetal forces at work in this prototypical Web 2.0 genre.

In the next chapter, authored by ELISA CORINO and CRISTINA ONESTI, attention narrows down on a specific type of Web application, the Newsgroup, the main object of investigation being agreement and disagreement routines within this new environment. In the Newsgroup, where dense interactional patterns take place and many people are involved, even agreeing and disagreeing strategies seem to follow well-defined schemes and rules. Such strategies are analysed in a qualitative survey aiming to be as representative as possible of the whole agreement/disagreement scale. After defining the characteristics of the newsgroup as a genre, the authors also discuss the use of *quoting* devices, given that messages in a newsgroup show a constant resumption of the original text of previous messages or parts of them, like self-sufficient microtexts.

In the following chapter, authored by ENRICO GRAZZI, the focus shifts to the use of web-mediated applications in language learning, and notably to the pedagogical implications of using English as a Lingua Franca at school in Internet-mediated communication. In particular, Grazzi discusses the use of Internet memes, user-generated verbo-pictorial metaphors exchanged by netizens, which represent a unique semiotic phenomenon peculiar to informal online communication. He argues that, since memes have become a typical feature of the informal language used by social networkers (especially younger ones), English teachers could exploit this resource to make students perform innovative language activities, conducive to turning them into members of a wider cyber community.

The third section of the volume focuses in particular on the blog, a Web 2.0 genre that has evolved rapidly, giving rise to a multiplicity of discursive phenomena, some of which have generic significance and seem to be on their way to consolidating into genres or subgenres in their own right. As the relevant literature has raised a number of questions concerning its status, the blog qualifies as one of the textual realizations that are most suitable to be used as case studies for the discussion of the theoretical issues involved in the adaptation of traditional analytical tools to Web-mediated genres.

The section opens with a study by GIULIANA GARZONE, which takes up these issues and uses the blog as a case in point to deal with the question of web genres' antecedents and of the mechanisms at work in the rise and evolution of web-native genres. In this respect, the blog is particularly interesting as, thanks to its peculiar properties that make it an attractive sociotechnical format for a wide range of purposes, it has developed into a variety of forms, in many respects profoundly different from each other, so much so as to cast doubts on its generic integrity. Analysing in particular two different kinds of blog, the news blog and the corporate blog, Garzone shows that they have a core of features in common, exhibiting superficial and peripheral formal traits which make them promptly recognisable and maintaining at least partially a personal/existential component. This confirms that the blog ultimately derives from the diary/personal journal genre and other related genres, but has hybridized with, or has been colonized by, other discourse genres customarily used in the relevant domains. In light of these considerations, Garzone sees the weblog as a (macro)genre which is used in various contexts of situation for a whole range of purposes, mostly different from its original ones.

The next study, by SANDRA CAMPAGNA, focuses on participatory journalism (one among many manifestations of the Web 2.0 collaborative agenda) and more specifically on the language of readers commenting on *The Economist* articles. What is at issue is the interplay between the commenters' community and the editor, and the relevant repercussions on linguistic/rhetorical styles as a result of this dynamics. Campagna analyses a selected sample of comments responding to an *Economist* leader, and in particular tackles the controversial issue of the blog's generic integrity, which according to her in this context can hardly be determined on the basis of recurrent formal features, because of the constraints exerted by editorial imprinting. This leads Campagna to draw conclusions that are significantly different from those reached by Garzone in the previous chapter, as in her view the ephemeral quality of genres in web-mediated communication lends support to the unconventional definition of blogs proposed by Miller/Shepherd as "a technology, a medium, a constellation of affordances – and not a genre" (Miller/Shepherd 2009: 283), which therefore in her view deserves further consideration.

The discussion of the blog continues in the next chapter, authored by MAŁGORZATA SOKÓŁ which marks a shift from general issues to the examination of more specific aspects, as it focuses on authoring practices in academic weblogs, looking at authors' self-presentation through the use of metadiscoursal resources of self-mention and citation. This type of analysis makes it possible to describe how authors explicitly mark their presence in weblog entries through self-reference, and how they strengthen their credibility by referring to reliable sources of information. On the one hand, the chapter's findings present academic weblogs as platforms for informal self-expression and self-promotion under the generally accepted, 'traditional' norms of scholarly exchange. On the other hand, the research indicates the emergence of less constrained modes of academic communication that result from the current pursuit of 'open science', where blogs have a potential to bring science to the public.

The field of investigation of the last two chapters in the volume is a relatively recent evolution of blogging, microblogging, which is now developing into a separate and autonomous area of the blogosphere. Quite interestingly, both studies discuss the use of microblogging in corporate communication, a fact that highlights the flexibility of this web genre and at the same time emphasizes the keen interest of the corporate world, and in particular of marketing managers, for new social media.

In the first of the two chapters, GIORGIA RIBONI starts from the observation that communication through social networks and microblogs allows big corporations to appear more 'human' in their customers' eyes. Combining recent research on web genres and discourse analytical tools, she studies the participation framework characterising multinational corporations' communication with potential buyers through microblogging, and the relevant marketing strategies. Drawing on data collected on the Twitter profile of the *Whole Foods Market* (an organic food multinational based in Texas), she discusses how Erving Goffman's notions of 'animator', 'author', 'principal' as well as 'ratified' and 'unratified participant' (Goffman 1981: 144) can be respectively applied to corporate microbloggers and customers/navigators. The analysis highlights that *Whole Foods* twitterers tactically vary the participation framework in their interaction with users in order to be able to better promote their company's products.

The book closes with a study by MARIA CRISTINA PAGANONI which presents recourse to social networks as a cogent change in advertising, now increasingly integrated into corporate websites, which reflects later trends in consumers' behaviour. The ways in which Twitter has rapidly become a key tool in a company's branding strategy is illustrated by analyzing its multiple functions in the case of a major international brand (Coca-Cola). It is argued that new genres like tweets do not supersede previous advertising genres, but rather co-exist side by side with them, borrowing from them and enhancing their mutual role, while maintaining their distinctive traits. It is Paganoni's contention that although Twitter's conciseness may appear problematic, the oversimplification of linguistic features and rhetorical strategies it requires can be turned into a communicative advantage, as tweets are highly effective in customizing promotional messages that are richly in tune with the underlying corporate philosophy.

Overall the picture sketched by the studies collected in this volume confirms the important role that web-mediated communication plays in all walks of life today. The WWW has come of age, so to speak, and we are now capable of exploiting its potentialities much more extensively and systematically than in its early stages. Its impact has been not only interactional but also transactional; in other words it has not been merely virtual, as popular wisdom has it, but it has radically transformed the way we get things done, profoundly modifying our daily life in its civil, economic, social and political aspects with important implications in terms of ease, rapidity, effectiveness, as well as pervasiveness, the latter characteristic also having the effect of turning it into an important instrument of democratization in many fields (cf. Margolis/Moreno-Riaño 2009). With Web 2.0, new social spaces have been put at our disposal, and this brings also an exquisitely individual, personal dimension to the use of technologies. Within this picture, social practices have evolved with unimaginable rapidity, and have often taken totally unexpected directions. This has caused genres to change as rapidly and unpredictably, also adapting to uses that would not have been conceivable only a few years ago, and determined the birth of more genres, new in shapes and modes.

Hence the challenge for the discourse analyst to describe and account for phenomena that are inherently proteiform, and inhabit a world which is partially parallel to the material space where our life is set, but at the same time has interesting interferences with it, renewing it in some of its very basic coordinates. Inevitably, this determines an extremely dynamic profile for the genre-oriented area of discourse studies, where not only descriptions but also analytical tools require incessant updating and finetuning.

References

Archibald, James 2007. Freedom of Expression, Opinion and the Press in the Gulf States: Weblogs and Emergent International and Regional Media Policy. In Garzone, Giuliana / Catenaccio, Paola / Poncini, Gina (eds) 2007. *Multimodality in Corporate Communication. Web Genres and Discursive Identity*. Milano: FrancoAngeli, 167-179.

Askehave, Inger / Ellerup Nielsen, Anne 2004. *Webmediated Genres – A Challenge to Traditional Genre Theory. Working Paper nr. 6.* Aarhus: Center for Virksomhedskommunication.

Askehave, Inger / Ellerup Nielsen, Anne 2005. What are the Characteristics of Digital Genres? – Genre Theory from a Multi-modal Perspective. *Information Technology & People*. 18/2, 120-141.

Askehave, Inger / Swales, John M. 2001. Genre Identification and Communicative Purpose: a Problem and a Possible Solution. *Applied Linguistics*. 23/2, 195-212.

Barthes, Roland 1970/1974. *S/Z* (transl. by R. Miller), New York, Hill and Wang (*S/Z*, Paris, Éditions du Seuil, 1970).

Bhatia, Vijay K. 1993. *Analysing Genre: Language Use in Professional Settings*. Harlow/New York: Longman.

Derrida, Jacques 1972. *La dissémination*, Seuil, Paris.

Fairclough, Norman 1995. *Media Discourse*. London/New York: Arnold.

Friedman, Thomas L. 2005. *The World is Flat*. London: Penguin Books.

Garzone, Giuliana 1997. Osservazioni sul valore pedagogico dell'ipertesto. In Schena, Leo / Garzone, Giuliana (eds) *Il computer al servizio degli insegnamenti linguistici, Milano 28.4.1995-3.5.1996*. Milan: Centro Linguistico Università Bocconi, 21-39.

Garzone, Giuliana 2002. Describing e-commerce communication. Which models and categories for text analysis?. In Evangelisti, Paola / Ventola, Eija (eds) *TEXTUS* (*English in Academic and Professional Settings: Techniques of Description / Pedagogical Application*). XIV/2, 279-296.

Garzone, Giuliana 2007. Genres, Multimodality and the World-Wide Web: Theoretical Issues. In Garzone, Giuliana / Catenaccio,

Paola / Poncini, Gina (eds) 2007. *Multimodality in Corporate Communication. Web Genres and Discursive Identity*. Milano: FrancoAngeli, 15-30.

Glanz K., Rimer B.K. and Lewis, F.M. 2002. *Health behavior and health education: Theory, research, and practice*. San Francisco: Jossey-Bass.

Goffman, Erving. 1981. *Forms of Talk*. Philadelphia: University of Pennsylvania Press.

Hoffman, Donna L. / Novak, Tom P. 1996. Marketing in Hypermedia Computer-Mediated Environments: Conceptual Foundations. *Journal of Marketing* 60/3 (July), 50-68.

Landow, George P. 1992. *Hypertext: The Convergence of Contemporary Critical Theory and Technology*. Baltimore: Johns Hopkins University Press.

Lemke, Jay L. 2002. Travels in Hypermodality. *Visual Communication* 1/3, 299-325.

Margolis, Michael / Moreno-Riaño, Gerson 2009. *The Prospect of Internet Democracy*. Farnham: Ashgate.

Miller, Carolyn R. / Shepherd, Dawn 2009. Questions for Genre Theory from the Blogosphere. In Giltrow, Janet / Stein, Dieter (eds) *Genres in the Internet: Issues on the Theory of Genres*. Amsterdam: John Benjamins, 263-290.

Sosnoski, James 1999. Configuring as a Mode of Rhetorical Analysis. In Jones, Steven (ed.) *Critical Issues and Methods for Examining the Net*. Thousand Oaks (CA): Sage, 127-143.

Stöckl, Hartmut 2004. In between modes. Language and image in printed media. In Ventola, Eija / Cassily, Charles / Kaltenbacher, Martin (eds) (2004) *Perspectives on Multimodality*, Amsterdam/Philadelphia, John Benjamins, 9-30.

Swales, John M. 1990. *Genre Analysis: English in Academic and Research Settings*. Cambridge: Cambridge University Press.

Web 1.0

PAOLA CATENACCIO

A Genre-Theory Approach to the Website: Some Preliminary Considerations

1. Research aim and study design

It is an unquestioned fact that contemporary communication practices have been greatly affected by the rapid technological developments of the last 30 years or so, which have led – amongst other things – to the creation and spread of the World Wide Web as a privileged means for the storage, communication and retrieval of information.

The web environment undoubtedly offers novel and unprecedented affordances for the presentation of all kinds of information. In the greatest majority of cases, such information is made available within websites, which can be considered *prima facie* as collections of written texts and other oral/visual documents created and maintained by organisations or individuals to communicate information about themselves to the public at large. Websites come in many shapes and forms; indeed, because of the great variety of websites existing on the web, the word 'website' can be considered as a sort of umbrella term encompassing a huge variety of different entities. Despite its internal variability, however, the website has a psychological reality as a clearly identifiable object – or indeed a class of objects – sharing standard characteristics which web-literate users have become extremely adept at negotiating and navigating.

This blend of structural predictability (albeit combined with a high degree of flexibility) and content variability makes the website an uneasy concept to deal with. Despite its dominance in contemporary communication practices, little has been done to date to provide a systematic description of the website applicable to all instances of the construct, and capable of accounting in theoretically motivated terms for its status as both a semiotic artefact in itself, and an 'empty' template which can be filled with the most disparate information, giving rise to distinctly different objects.

But what is a website? Is it a medium, a technology, or a genre? Where is it located along the continuum between technological support and rhetorically motivated structure? Is it simply a container into which different contents can be poured, or does its structure constrain the selection of allowable contents, especially for what concerns their mutual relationships? Finally, and at a more abstract level, can its status be explained in a genre-theory perspective – and, conversely, can genre theory benefit from the exploration of the issue? This chapter attempts to answer these questions, taking theoretical accounts of emerging multimodal genres as its starting point and gradually broadening the perspective to encompass the website both as a self-standing construct and as a 'container' for a multiplicity of mutually interconnected but autonomous documents.

To pursue the above-mentioned aim, a methodological approach steeped in genre analysis has been adopted. More specifically, studies of genre analysis in the rhetorical tradition will be relied on, with the additional contribution of insights drawn from current literature on genre classification conducted within the field of computer science. Special attention will be devoted to existing conceptualisations of genre clusters, with the aim to verify whether they can be applied to the collections of pages and documents which make up a website.

The first step in the research will be a short critical review of the state-of-the-art literature on cyber/webgenres in the rhetorical tradition, aimed at taking stock of the insights provided and at outlining its impact on genre theory evolution. Secondly, the challenges posed by the application of principles drawn from genre theory to an entity such as the website, with its rhetorically underdetermined status, will be discussed. Next, a brief outline of some key aspects of a specific website – the Siemens global website, <www.siemens.com> – will be given, with a view to providing empirical evidence for the considerations made in the foregoing discussion. Finally, an attempt will be made to formulate a theoretically sound definition of the website.

2. Genres and the web: theoretical and empirical approaches

As hinted above, research on web-mediated communication has increased exponentially over the last few years, no doubt in consequence of the rise of the Internet as a privileged medium for the transmission of information. The existing literature covers a variety of scholarly approaches, ranging from multimedia literacy, to hypermedia research (and more specifically computer and information science), to discourse analysis (cf. Djonov 2005) and beyond.

The present study takes a genre-theory approach to the systematic description of the website as a construct, a choice which firmly anchors it within the overarching perspective of discourse analysis. So, while insights from other disciplines will be occasionally referred to, the main perspective remains rhetorical in nature. Of the other approaches mentioned above, the most relevant for the purposes of this study is to be found in genre-related studies conducted within the domain of computer science. These studies are mostly concerned with the application of genre-theoretical principles to web document design, identification or retrieval, or to the analysis of web-mediated communicative activities, and are therefore empirical in nature. In this strand of research, genre-theoretical principles typically provide a framework within which empirical research can be located, thus representing an excellent example of the suitability of discourse theories to frame research objectives in other domains.

With reference to the relationship between genre theory and computer systems research, it has to be pointed out that the influence trajectory goes for the most part from the former to the latter, in the sense that research in computer systems takes the genre concept wholesale from genre theory, albeit with subsequent adaptations to suit the requirements of the environment within which it is to be applied. There are also instances, however, in which the influence is reversed, and computer systems principles are brought to bear on genre theory. This is the case, for example, of research carried out by Yates and Orlikowski and their group (Yates/Orlikowski 1992, 2002; Orlikowski/Yates 1994;

Yates/Orlikowski/Renneker 1997; Yates/Orlikowski/Okamura 1999), who have concentrated on studying groups and teams adopting computer-mediated technology for collaborative work, and on analysing the consequences of computer-mediated communication practices on genres and repertoires of genres available to professionals operating in institutional contexts. Owing to the dominance of web-mediated communication in the contemporary world, the inclusion of technological factors in the development of genre theory is bound to increase in the future, as highlighted by some of the most recent literature published within a genre-theory framework (cf. amongst others Garzone 2007), and a more balanced relationship between the two traditions is likely to develop.

2.1. Genres, the web, and challenges to genre theory

During the last few years research in genre theory and analysis has repeatedly engaged with the difficult task of developing methods capable of accounting for the plethora of new and 'migrated' (i.e. transferred from print to the digital medium; cf. Crowston/Williams 2000; Shepherd/Watters 1998) genres that have arisen and/or developed within the hypertextual and multimodal environment of the World Wide Web. Indeed, the advent of the WWW with its affordances and flexible and multifarious environment has brought about not only new genres, but new challenges for genre theory (cf. Askehave/Ellerup Nielsen 2005; Lemke 2005; Garzone 2007). Hypermodality – probably the most salient feature of the web, which Lemke (2002a: 301), building on Kress/van Leeuwen's (1996) definition of multimodality, describes as the "conflation of multimodality and hypertextuality" – demands that the plethora of semiotic objects involved in web-based communication be considered not independently of one another, but in their interactions, as "signifiers on different scales of syntagmatic organization [...] linked in complex networks or webs" (Lemke 2002a: 300).

Such networks or webs come to represent what Lemke (2003, 2009) refers to as *traversals*, i.e., in Wodak's restatement of the concept (2006: 3), "virtual genres which transcend the traditional norms of time and space in specific, fragmented ways, constructing hybrid patterns of coherence". In the virtual space of the web, traversals not only bring different communication modes (written text, images, sound) to bear upon one and the same communicative event, but also represent linkages among different genres (a journalistic article can link to an interview, the interview to background information such as biographical notes, timelines of events, reports and so on), which together contribute, in variable sequencing, to the realisation of the event itself.

The theory and practice of genre analysis has been deeply affected by the above-mentioned evolution. Having to account for multimodal, hypertextually organised materials has played a crucial role in the process of reconceptualisation of the very notion of genre. The distinction between reading and navigating modes, and the need to account for the latter, has of course been crucial. As Lemke (2002b: 52) points out, in a hypertextual environment "one tries to offer readers/users opportunities to make meanings of their own, to come to their own conclusions, based on the web of related elements provided, which may be combined and sequenced, logically, temporally, or experientially in many different ways". But in many cases the "related elements provided" are not 'simple' semiotic objects; in fact, they often are self-sufficient texts belonging to recognisable genres, which, therefore, come to play and interact in the broader semiotic space inhabited by the communicative event as a whole.

This latter aspect brings to the fore another important issue which has been high on the agenda of genre theorists and analysts over the last few years. The synergic interplay among different genres which typically characterises hypermodality demands a shift in focus from the analysis of individual genres to the investigation of the *relationships* which exist among genres involved in hypermodal communication activities. As Swales (2004: 5) points out,

> while a decade or more ago it might have been justifiable [...] to focus on gaining a better understanding of *single* genres [...] in the very early twenty-first century [...] concentration on such partial, if often valuable, studies is no longer the only possible approach.

Indeed, over this period some of the most innovative research in genre theory has focused, on the one hand, on the relationships which hold among genres, and, on the other, on the way the integrity of each genre is challenged (but at the same time creatively stimulated) by exchanges of linguistic and rhetorical resources, with concepts such as genre mixing and genre embedding featuring high on the analytical paradigms of genre theorists (Bhatia 1995; 1997; 2004).

As this brief outline suggests, recent advances in genre theory provide valuable insights that can be fruitfully applied to an analysis of the website which takes stock both of its perceived self-contained (though potentially open-ended) nature, and of its function as a crossroads of interrelated texts – "a medium and a technology which encourages and affords ease of constructing sequences of textual units that are not uniquely determined by, or even in many cases anticipated by, the authors or designers of a particular hypertext web" (Lemke 2005: 52) and which is capable of providing a preliminary account of the nature and structuring of the website as a collection of hypertext genres.

2.2. Genre in computer systems research

The genre construct has also been fruitfully applied to computer systems research. Its relevance for this field can be easily surmised from the following quote, which is taken from the introduction to *Genres on the Web*, a recent edited volume (Mehler/Sharoff/Santini 2010) gathering contributions to the study of genre in a system science perspective:

> But why is genre important? The short answer is: because it reduces the cognitive load by triggering expectations through a number of conventions. [...] Essentially, knowing the genre to which a text belongs leads to predictions concerning form, function and context of communication. All these properties together define what Bateman calls the "the most important theoretical property" of genre for empirical study, namely the power of predictivity. The potential of predictivity is certainly highly attractive when the task is to come to terms with the overwhelming mass of information available on the web. (Santini/Mehler/Sharoff 2010: 4)

Studies of webgenres, or, following Shepherd/Watters (1998), cybergenres, have borrowed from the rhetoric-based notion of genre (mostly adopted via Orlikowski/Yates 1994) the key concept of functionality, thereby introducing a new criterion for the classification of documents which has proved extremely influential in the development of methods for the automatic classification of webdocuments (cf. Watters/Shepherd 1999; Shepherd/Watters/Kennedy 2004; Santini 2004, 2005; Kennedy/Shepherd 2005). But more importantly for the purpose of this study, research on genres on the web has highlighted that genres frequently have composite and multi-layered structures which often tend to defy univocal classification. In Crowston and Kwasnik's (2004: 3) words,

> many genres (such as a newsletter, for instance) can be viewed as composites of several genres (articles, editorials, calendars of events) and can be distinguished by both their components as well as by the unique assembly of components into an identifiable whole. From a classification point of view, this means establishing a scale for the scheme. How finely grained does the identification of genres (and their possible components) have to be? Conversely, how do we know when we have reached the boundaries of a given genre? When does a memo turn into a report or an abstract into a review?

Existing taxonomies span from extremely fine- (cf. Crowston/Williams 2000) to very coarse-grained (Watters/Shepherd 1999; Watters/Dong 2006; Lindemann/Littig 2011), often using websites as the unit of analysis. This, however, is rather problematic, because websites are typically polymorphic (i.e. they come in many shapes and forms, even when sharing essentially similar communicative purposes; cf. Mehler/Gleim/Wegner 2007), and by definition comprise more than one type of text, each of them potentially belonging to a different genre. Thus, even while being important in their own right, taxonomies are hardly suited to providing a theoretically motivated explanation of the website concept, which in the majority of cases remains out of focus. A partial exception is represented by a recent study by Lindemann and Littig (2011), where website structure features as a variable in the classification criteria adopted, but – to the best of my knowledge – no systematic analysis has been carried out yet. These considerations suggest that the website has so far been a heuristically useful, but theoretically underdetermined starting point for the investigation of web-mediated genres – a fact that makes a systematic account of this construct ever more necessary if its centrality in system science genre-based research is to be justified.

3. The website between technological affordances and rhetorical function

3.1. Medium, genre, and rhetorical function

As mentioned above, providing a theoretically-grounded definition of the website entails a dual perspective on the website as both a whole and a multi-genre container. As regards the website as a self-standing entity, it has to be pointed out that much recent research has focused on specific types of website, rather than on *the* website as such. In particular, it has been argued that websites can be categorised as belonging to a limited number of 'super-genres' (cf. Lindemann/Littig 2011). This view is grounded in the recognition that it is fairly easy to distinguish a corporate website from an academic or a commercial one, a fact that is linked to each (super)genre's ability to raise genre-specific rhetorical expectations; yet, at the same time, there are also rhetorical expectations which are not dependent on the single genre, but are common to all of them by virtue of their being websites.

It has been maintained that a categorisation of the website as a *medium* can account for such expectations. This would entail, following Lemke (2002b), considering the website as a hypertext, endowed with properties "analogous to the functional properties of a book, as opposed to either its textual contents or its physical make-up". However, a website is not just a hypertext, but rather a *type* of hypertext organised in a recognisable – and above all predictable, even within the above-mentioned variability – manner. A hypertext as such does not raise *rhetorical* expectations, although it may raise expectations in respect of the 'hypertext activities' that may be carried out in and through it (Lemke 2002b). A website, by contrast, does (cf. Symonenko 2007a). A first glance at the homepage immediately raises genre-related expectations, not only in respect of the page, but in respect of the entire website, and makes it possible to identify potential navigation paths leading to the required information, thereby creating traversals across documents and/or document parts which ultimately constitute genre-like structures (Lemke 2005). This suggests that besides being a (sub)genre in itself (indeed, possibly the first uniquely digital genre; cf. Dillon/Gushrowski 2000), a website homepage is also part of a broader structure (the website) which it helps understand and navigate. Its function, in this sense, is not unlike that of the index page in a book, to carry on Lemke's metaphor, except that, of course, it has a dynamic nature, even as an index. But in moving from the functional properties of the medium to its content-related and therefore rhetorical features, as we do when we move from book, i.e., a medium, to index page, a functional attribute which potentially raises much more specific rhetorical expectations, we are already stepping over the boundaries of medium into the territory of rhetorical organisation and, therefore, genre.

Thus, while hypertext is the medium, and a specific website can be identified as a specimen of a recognisable and describable genre, the website, disembodied, so to speak, from its specific contents, can be seen as a type of hypertext located in a sort of intermediate position between medium and genre. Evidence of the difference between the hypertext medium and the website can be found, for instance, in the fact that whereas hypertext is characterised by a lack of clear boundaries (Snyder 1997: 127), websites can be (and often are) hypertextually linked to documents which fall beyond their boundaries, but these boundaries are still recognisable, albeit porous. Thus, a better definition of the website as a concept could perhaps be that of *rhetorical interface*. The website construct mediates between the affordances of the medium and the rhetorical function(s) of the genre, providing the structural features required to deploy them.

The concept of rhetorical interface provides a framework for interpreting the website as an entity having an identity and boundaries of its own (however permeable the latter may be). At the same time, it is also a flexible enough concept to accommodate a conceptualisation of each individual website as a *collection* of genres. This is not in contrast with the fact that each individual website is also a specimen of a single identifiable genre. The possibility of embedding multiple genres within a broader one has long been recognised, and indeed, as highlighted above (§ 2.2), "many genres […] can be viewed as composites of several genres" (Crowston/Kwasnik 2004: 3). Moreover, insofar as websites can be seen as structures which enable the accomplishment of different tasks, all of which are subservient to a set of overarching institutional/organizational aims, it is mandatory to view the different genres which are used to accomplish them as interrelated. This entails a consideration of the various genres which make up a website as an organised cluster.

3.2. Genre clusters and the website

The notion of genres being interrelated so as to form broader structures (or assemblages, as Spinuzzi [2004] refers to them) is not new, and several denominations have been proposed to explain the various relationships which hold among genres within each cluster. The one which has been favoured within the field of computer studies is that of *genre system* (although more recently it has been partly supplanted by that of *genre ecology*), which was first coined by Bazerman in 1994 and later picked up by Yates and Orlikowski (2002). Bazerman proposed the term *genre system* to refer to interrelated genres that interact with each other in specific settings, e.g. an application and a patent; Yates and Orlikowski further specified the concept, applying it to collections of interdependent genres typically enacted in some conventional sequence (or limited set of acceptable sequences) in relation to each other, whose purpose and form interlock and whose combined implementation is in the service of organisational functioning.

Besides that of genre system, other conceptualizations of genre clusters exist. Among them, the best known ones are those of *genre set* (Devitt 1991), *genre repertoire* (Orlikowski/Yates 1994), *genre ecology* (Freedman/Smart 1997; Spinuzzi/Zachry 2000; Erickson 2000; Spinuzzi 2003), *genre network* (Swales 2004), as well as *genre colony* (Bhatia 2002) and *genre chain* (Swales 2004; Fairclough 2004).[1] Insofar as they focus on genre interrelatedness, all these conceptualisations share some features, but there are also distinctive elements which set each of them apart from the others.

---

1 The concept of *genre colony* is too broad to be useful for the purposes of this study, whereas that of *genre chain* limits too narrowly the relationship between genres to one of sequentiality.

The notion of *genre set* was first proposed by Devitt in a seminal paper on intertextuality in tax accounting (Devitt 1991). In this paper Devitt argued that in carrying out their activities accountants rely on genre sets, i.e. assemblages of genres through which their professional tasks are accomplished. Compared to this definition, which focuses on a specific professional community, the concept of genre system proposed by Bazerman (1994) is broader, as it comprises in its scope not only the genres used by a specific professional group, but the entire cluster of genres used by all participants in institutional exchanges to perform their communication activities.

The idea of structuration is equally crucial to the concept of *genre repertoire* developed by Orlikowski and Yates (1994). A genre repertoire differs from a genre set or a genre system in that it does not refer to a sequentially organised cluster of genres, but rather to a group of partially overlapping genres, i.e. genres which can be used interchangeably and/or sequentially (to an extent at least) to perform a given organisational function. As a result, the concept of genre repertoire is more flexible (and therefore has greater explanatory validity) than that of genre system, with its pre-ordained sequential structure, but is nonetheless strictly linked to institutionally established practices which are performed through genres.

The notion of *genre ecology* is also linked to the concepts discussed above, but takes an even more radical approach to the relationship between genres and institutions. As succinctly put by Spinuzzi (2004: 6),

> genre ecologies are grounded in theories of distributed cognition, particularly activity theory, and consequently emphasize genres as collective achievements that act just as much as they are acted upon. Whereas the other frameworks are firmly asymmetrical, emphasizing a human being's control over, performance of, and communication through genres, the genre ecology framework is more symmetrical, replacing the notions of performance and communication with the notion of mediation.

Supporters of the genre ecology concept argue that this construct is more suited to explaining the way in which institutions function interactively through the mediation of genres, as they shift part of the agency from genre users to the genres themselves. They maintain that doing so provides a better theoretical framework for accounting for the way in which the patterns of distributed cognition which characterise institutions are developed and established over and across both official and unofficial (sometimes occluded; cf. Swales 1996) genres, and for the processes of genre hybridisation, change and new creation which distinguish dynamic genre scenarios such as web-mediated ones.

Despite their different emphases and peculiarities, all four concepts described so far share a common focus on the way in which genre-mediated activities are brought to bear on the functioning of organisations, either sequentially or simultaneously. They presuppose that such activities can be identified and classified, and their interconnections mapped in fairly predictable – although dynamic – ways. They also inscribe the investigation of genre-mediated activities within institutional boundaries, and assume that participation frameworks in the institution-based communication activities examined can be identified.

But does all this apply to the way in which communication is mediated within websites and throughout the different genres which make up each and every website? Symonenko (2007b: 1) starts from the assumption that this is the case. She states that

> a website [is] an integrated information package created by the site sponsor with a particular audience in mind (for example a university website is geared towards prospective and current students, faculty, alumni, potential sponsors with an interest in research etc.). Collectively, the sponsor and the audience represent the site's discourse community, which, according to genre theory [...], shapes the site as a communication artefact, in terms of the kind of content presented there, as well as the way it is presented.

While websites are obviously organisational/institutional artefacts, the notion of discourse community as interpreted by Symonenko is, I believe, rather misleading. Websites are undoubtedly designed with a set of audiences in mind, but their publication on the World Wide Web broadens the participation framework to the point of making it virtually unpredictable. Reading paths for each different envisaged audience may be partially inscribed in the website design, but strict constraints go counter to the hypermedia nature of the website, and cannot be enforced. When an attempt is made to narrow down the navigation possibilities available to potential users (for instance by limiting the functional properties of the website), the consequence is usually a reduced exploitation of the affordances of the medium – a fact which is typically interpreted as a design failure. Well-designed websites, by contrast, are amenable to multiple navigation options giving rise to a variety of reading paths involving different genres, or indeed involving the same genres, but for different purposes. The reading path of a potential investor in a private university may have very little in common – if anything at all – with that of a student. And the sustainability reporting section of a corporate website may be read with very different purposes in mind by investors and environmental activists, thereby contributing to structuring altogether different communication activities.

Accounting for the way in which genres are mutually connected within a website, thus, calls for a consideration of both institutional activities and purposes as they are mediated through genres, and of what individual users make of the affordances, properties and purposes such genres provide and/or convey. Swales's notion of *genre network* (2004) may be useful in this quest for systemic accountability combined with user-dependent flexibility. Swales (2004: 23) defines a genre network as "the totality of genres available for a particular sector [...] as seen from any chosen synchronic moment". He rejects the term 'system' on the grounds that it "suggests that we have a greater understanding of how everything fits together in a 'system' than is likely the case". By contrast, he maintains,

> networks operate at an intermediate level between the structural properties of institutions (such as those reflected in an organizational chart) and the communicative activities of single individuals. Having a network frame allows us, by tracing intertextual links and other kinds of recontextualization, to place individual genres within a heuristically valuable wider context.

By looking at the website as a network of genres which are interconnected by virtue of the mediating role they have between institutions and individuals it is possible to account for both the totality of the genres represented, and their individual functions. Moreover, by keeping the relationship between genres flexible enough to allow for different recontextualisations of each and the same genre within a variety of navigation paths leading to the generation of different traversals (as per Lemke's definition quoted above, cf. § 2.1), a conceptualisation of this kind can account for the different uses to which texts and genres are put by readers pursuing different agendas. According to Linell's (1998: 14) classical definition, recontextualisation "involves the extrication of some part or aspect from a text or discourse, or from a genre of texts or discourses, and the fitting of this part or aspect into another context, i.e. another text or discourse (or discourse genre) and its use and environment". If we extend this definition from text to genre, we can see how the different genres which make up a website can be recontextualised – sequentially, even while remaining within the boundaries of institutionally sanctioned interpretations, or discursively, thanks to the possibility afforded by the web to create meaning *across* intra- and inter-institutional boundaries.

The possibilities for recontextualisation are indeed multiplied by the hypertext medium, which fosters a dynamic approach to text-building. As Lemke (2005: 52) highlights,

> hypertext [...] is a medium and a technology which encourages and affords ease of constructing sequences of textual units which are not uniquely determined, or even in many case anticipated by designers of a particular hypertext web. [...] [Authors/designers try] to offer readers/users opportunities to make meanings of their own, to come to their own conclusions, based on the web of related elements provided which may be combined and sequenced, logically, temporally, or experientially in many different ways.

Such text building processes are trans-generic (Lemke 2005: 51), and appear to be based on relevance relations (cf. Rauen 2009, quoting Blass 1990), as opposed to sequentially cohesive chains. Accessing genres through hypertext, thus, involves recontextualising them in light of individual agendas which can only partially be predicted or constrained. The network metaphor captures the dynamic interconnectivity of genres which not only mediate institutional activities, but can also be used creatively to generate alternative meanings in unpredictable – and potentially idiosyncratic – ways.

To conclude, it is my contention that the website can be described as a hypertextually organised *rhetorical interface* providing structured access to a *network of genres* which – in varying sequencing and through multiple processes of recontextualisation – enable the creation of user-generated traversals that are nonetheless at least partially institutionally constrained, albeit only loosely so.

4. An empirical example: the Siemens website

In order to ground the claims made in the previous paragraph in empirical evidence, in this section a brief analysis will be conducted of a subsection of the Siemens corporate website. Starting from the corporate homepage, a description will be provided of selected structural aspects of the website, and a tentative account will be given of the intertextual and inter- and trans-generic links between the various documents accessible from the website itself.

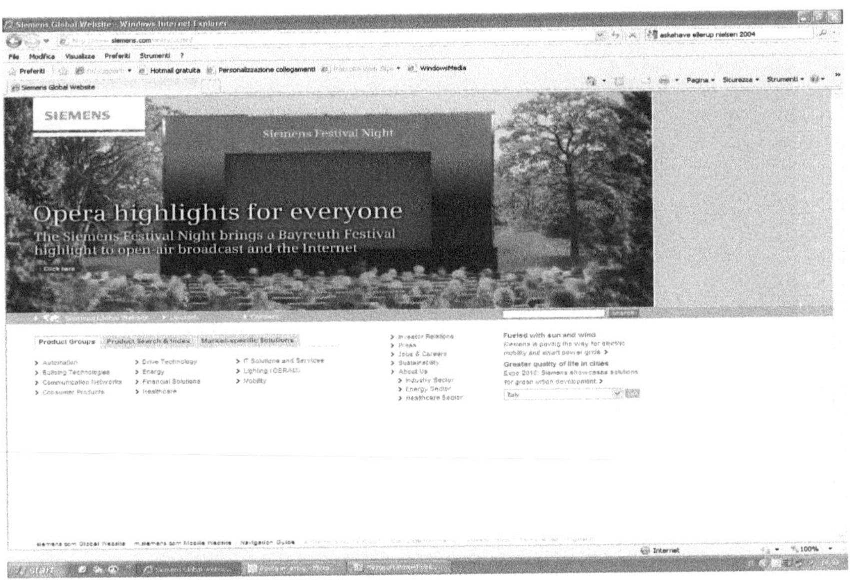

Figure 1. Siemens homepage, August 2010.

The Siemens homepage is rather unusual in its outlook in that the screenspace is almost entirely devoted to an event (the Bayreuth Op-era Festival) sponsored by the company. The picture and caption which occupy centrestage in the homepage link to a dedicated page (see fig. 2) which, in turn, branches off into a number of other pages, both within and without the boundaries of the Siemens website. The analysis which follows focuses not only on the general website, but also on the dedicated page itself, and on the subsection of the website it gives access to. This subsection appears to all effects to be self-standing, and therefore amenable to the same considerations that can be made in respect of the superordinate website. Moreover, by showing how certain principles can be applied to artefacts of varying size and complexity, the analysis can also illustrate that the principles outlined above can be applied recursively, thereby increasing their explanatory capacity.

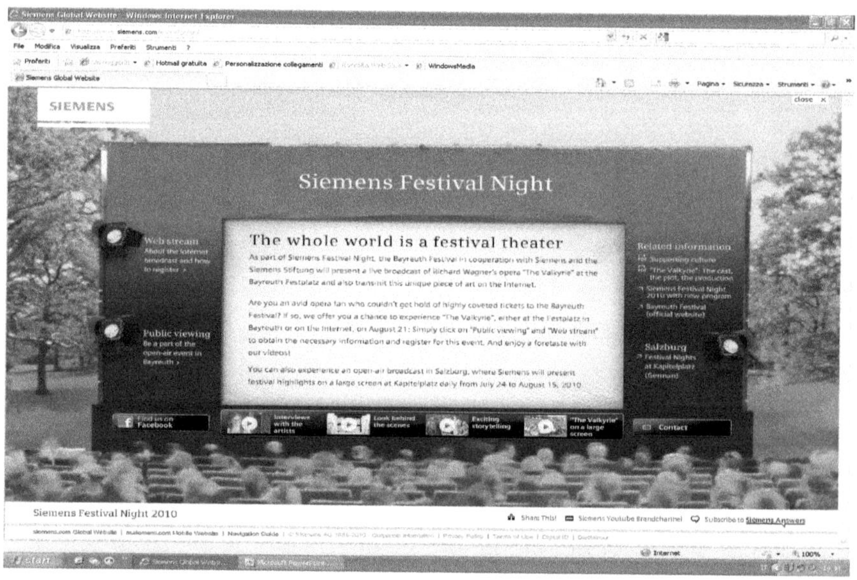

Figure 2. Siemens Festival Nights subsection within the Siemens website.

The 'Siemens Festival Night' page can be considered as a gateway to the event's dedicated website. Besides featuring links to the web-streaming of the event and to a public viewing platform, the page also has a related information section with links to pdf documents – one on how to support the festival, and another on the plot of *The Valkyrie*. Three more links lead respectively to a section dedicated to Siemens' involvement in the festival, to the festival's official website (external to the Siemens site), and to information on special festival nights to be held in Salzburg. At the bottom of the page are a number of links to multimedia materials related to the festival, as well as one to the sponsored event's FaceBook page, and a 'Contacts' section. Even a cursory glance at the materials accessible from the Siemens Festival Nights section indicates that while the section as a whole is identifiable as a distinctive and structurally self-contained unit, it also functions as a collection place for a variety of different genres whose individual peculiarities and affordances are brought to bear upon the overall aim of the section, which is to inform about the festival and highlight Siemens' contribution, thereby serving a more general self-promotional purpose.

The materials featured span from multimedia genres (both performances and documentaries about the performances), to re-mediated genres such as informative brochures in pdf format, to external websites with their own complex articulations and communication agendas. The sheer variety and self-standing nature of the genres included in the section suggests that each and every document is endowed with specific communicative aims, which are, however, hierarchically subsumed under an overarching institutional communicative purpose. This institutional purpose may not, however, be immediately apparent to website visitors interested in the sponsored event as such. Thus, the genres included in the section 'function' both independently, and as part of a broader communication activity, although users accessing the individual genres do not necessarily take part in the latter (a reader interested in the Festival Nights event may not be interested in Siemens' sponsoring agenda, having ended up on the corporate event page through a web search engine).

But how is this higher-level communication activity organised, and what relationship holds among the genres included in the website section considered? The key criterion here seems to be one of relevance: the materials included are all relevant to the main theme. However, individual documents do not need to be approached sequentially to be understood. The semiotic organisation of the event page suggests as much: links to the various genres appear to irradiate from the image occupying the centre of the screen, and are not hierarchically structured. In fact, a conscious attempt seems to have been made to organise them in a sort of 'satellite' manner, with no priority being given to any of them. Only the contacts and the Facebook links are given a clearly marginal position. Everything else is non-hierarchically, non-sequentially organised, thus suggesting an emphasis on user/reader *choice* in constructing individual traversals across genres.

If we now go back to the corporate homepage (Fig. 1), we can see that the bottom part of the screenview is occupied by links leading to different pages within the website, some of them being full-fledged genres (product descriptions, press releases, corporate overview, sustainability reports etc.). The links to further pages and genres are more clearly structured here, either thematically or hierarchically, but access to each of them is not sequentially constrained. Moreover, cross-referencing across pages is common (a press release on corporate responsibility will feature a link to the sustainability section of the site; description of energy-related products will also link back to sustainability texts; and the different genres included in the investors' section are not only intertextually connected, but also linked to other sections), thereby emphasising their interconnectedness.

Examples of this interconnectedness abound in any kind of website. With reference to the Siemens website, a clear instance of it can be seen in Example (1) below. The excerpt comes from a press release published on the Siemens website. The press release is not only featured on the dedicated newsroom page, but also in the bottom-part of the website section dedicated to Siemens Enterprise Communication. The text of the press release links back to the Enterprise page, and also to the specific product the text makes reference to (the links are underlined in the excerpt):

(1) Press Releases
Siemens Enterprise Communications Improves Efficiency and Mobility for SMBs with Updates to Industry-Leading OpenScape Office and HiPath 3000 UC Solutions
Nov 29, 2011
All-in-one UC suite includes enhanced mobility, tablet client, web collaboration capabilities and flexible deployment options

> Siemens Enterprise Communications today announced updates to its all-in-one unified communications and collaboration suite – improving mobility, increasing business efficiency and lowering costs for small to medium-sized businesses (SMBs).
> The upgrades to OpenScape Office and the HiPath 3000 voice platform include a new UC client for tablets, OpenScape Web Collaboration and a UC plug-in for Microsoft Outlook 2010. These solutions help SMBs better serve their customers and reduce communications costs – all among key priorities for SMB executives according to a just released report.

By providing these links, the company not only highlights, but in fact *constructs* intertextual and cognitively relevant relations even where they are not self evident. The sustainability section of the Siemens website, for instance, features a direct link not only to the company's sustainability report, but also to the annual report, thereby emphasising the complementarity of social and financial aspects of corporate performance. And while each of the two reports has a specific communicative aim, both include mutual references, as well as references to links where more information can be found on specific aspects of corporate performance. Thus, each genre is embedded in a network of texts, each of them not only providing additional information more or less closely related to the other texts, but also contributing to framing all co-referenced texts in a meaningful whole even while allowing for their independent status.

Communicative purpose seems therefore to be hierarchically organised, with individual genres pursuing rhetorical aims which are both an end in themselves, and at the same time serve higher purposes. Depending on their own needs and purposes, individual users are free to access any of the documents featured on the website, constructing meaning across genres and pages without following any preordained sequence, but taking advantage of the 'suggestions' provided in the website. Thus, while all genres and possible navigation routes jointly contribute to realising the overall set of communicative purposes of the company, the navigation paths of real readers can only realise some of the meaning inscribed in the site, or even generate new – and not envisaged – ones. Nonetheless, as mentioned above, the structural choices inscribed in the website (or in the various sub-websites) provide suggestions as to possible reading paths, hinting at, if not constraining, suitable traversals, and displaying recurrent cross referencing which highlight the interconnected (occasionally even circularly so) nature of the documents provided.

5. Conclusions

This chapter has attempted to provide an empirically grounded and theoretically motivated account of the website construct in a genre-theory perspective. To do so, recent developments in genre theory have been outlined, with special attention paid to those prompted by the need to account for new modes of textuality fostered by technological advances, in particular hypermedia. Parallel to this, consideration has also been given to genre-based studies of web-mediated communication conducted within the field of system science. The majority of these studies have an empirical focus, and rely on the genre construct as a working concept for the identification of functional as well as formal and structural regularities in web documents, mainly in the service of the development of automatic document retrieval methods.

The review of the existing literature has shown a remarkable lack of theoretical consideration for one of the most influential constructs in contemporary literacy – the website. While system science studies often seem to *assume* the genre status of individual websites, the website as a construct does not appear to have received due consideration in a genre-theory perspective.

In searching for a suitable definition of the website in a genre-theory perspective, the hypothesis that the website as a construct could be ascribed genre status has had to be discounted, as existing websites appear to be too markedly different to allow for a single overarching communicative purpose and rhetorical organisation. At the same time, it does not seem appropriate to define the website simply as a medium, since a medium is best described as a combination of affordances and constraints – not as a rhetorically organised whole. By contrast, the website construct seems to raise rhetorical expectations which have to be met for a specimen to be considered successful. Thus, a definition of the website as *rhetorical interface* between the hypertext medium and specific genres has been proposed.

The concept of rhetorical interface helps account for the perceived self-containedness of websites, however porous such self-containedness may be. At the same time, it is quite evident that websites have a composite nature, and that they embed a multiplicity of genres. These genres are interrelated, were it for no other reason than that they share a common virtual space within which they are endowed with separate but mutually accessible functions. But what is the nature of the relationship among these genres? A review of existing conceptualisations of genre clusters seems to suggest that many of them are steeped in institutional or professional environments characterised by fairly predictable participation frameworks. They are therefore lacking in the flexibility needed to account for unpredictable sequencing and associations which occur in website navigation, where institutions are forced to relinquish control over navigation paths, and users can construct meanings across genres and texts according to their interests or agendas. The only conceptualisation which seemed sufficiently supple to capture both the interconnectedness among the different genres accessible from a website and the multiple ways in which these could combine to generate meanings was Swales' (2004: 22) notion of *genre networks*. One aspect of genre networks which appeared to be particularly relevant for the application of the concept to the website was their theorised amenability to recontextualisation, as hypermedia textuality involves the creation of meaning across trans-generic traversals which tap into generic resources and recontextualise them in line with users' purposes and agenda.

Genre theory has only just started to engage with the challenges posed by hypermedia textuality. There certainly is a long way to go before satisfactory accounts are given of the way genres work in hypermedia environments. It seems likely, however, that notions like interface, network and recontextualisation will play a relevant role in such accounts, which will only be developed through the fruitful combination of theoretical and empirical approaches.

References

Askehave, Inger / Ellerup Nielsen, Anne 2005. Digital genres: A challenge to traditional genre theory. *Information Technology & People* 18/2, 120-141.

Bazerman, Charles 1994. Systems of genre and the enactment of social intentions. In Freedman, Aviva / Medway, Peter (eds) *Genre and the New Rhetoric*. London: Taylor and Francis, 79-101.

Bhatia, Vijay K. 1995. Genre-mixing in professional communication – The case of private intentions v. socially recognized purposes. In P. Bruthiaux, T. Boswood & B. Du-Babcock (eds), *Explorations in English for professional communication*. City University of Hong Kong, 1-19.

Bhatia, Vijay K. 1997. Genre mixing in academic introductions. *English for Specific Purposes*. 16/3, 181-195.

Bhatia, Vijay K. 2002. Applied genre analysis: Analytical advances and pedagogical procedures. In Johns, Ann M. (ed.), *Genre in the classroom: Multiple perspectives*. Mahwah, New Jersey: Lawrence Erlbaum Associates Publishers, 279-284.

Bhatia, Vijay K. 2004. *Worlds of Written Discourse*. London: Continuum.

Blass, Regina 1990. *Relevance Relations in Discourse*. Cambridge: Cambridge University Press.

Crowston, Kevin / Williams, Marie 2000. Reproduced and emergent genres of communication on the World Wide Web. *The Information Society*. 16/3, 201-215.

Crowston, Kevin / Kwasnik, Barbara 2004. A framework for creating a facetted classification for genres: Addressing issues of multidimensionality. *Proceedings of the 37th Hawaii International Conference on System Sciences*, Bie Island, Hawaii: IEEE.

Devitt, Amy J. 1991. Intertextuality in tax accounting: generic, referential and functional. In Bazerman, Charles / Paradis, Jane G. (eds). *Textual Dynamics of the Professions: Historical and Contemporary Studies of Writing in Professional Communities*. Madison, WI: University of Wisconsin Press, 336-357.

Dillon, Andrew / Gushrowski, Barbara 2000. Genres and the Web. Is the personal homepage the first uniquely digital genre? *Journal of the American Society for Information Science*. 51/2, 202-205.

Djonov, Emilia 2005. *Analysing the Organisation of Information in Websites: From Hypermedia Design to Systemic Functional Hypermedia Discourse Analysis*. Unpublished Doctoral Dissertation, The University of New South Wales. Available at <http://unsworks.unsw.edu.au/fapi/datastream/unsworks:901/SOURCE01>.

Erickson, Thomas 2000. Making sense of computer-mediated communication (CMC): Conversations as genres, CMC systems as genre ecologies. *Proceedings of the 33rd Hawaii International Conference on System Sciences (HICSS-33)*. Los Alamitos: IEEE Press.

Fairclough, Norman 2004. *Analyzing Discourse: Textual Analysis for Social Research*. London: Routledge.

Freedman, Aviva / Smart, Graham 1997. Navigating the current of economic policy: Written genres and the distribution of cognitive work at a financial institution. *Mind, Culture, and Activity*, 4/4, 238-255.

Garzone, Giuliana 2007. Genres, multimodality and the World Wide Web: theoretical issues. In Garzone, Giuliana / Poncini, Gina / Catenaccio, Paola (eds) 2007. *Multimodality in Corporate Communication. Web Genres and Discursive Identity*. Milano: FrancoAngeli, 15-30.

Kennedy, Alistair / Shepherd, Michael 2005. Automatic identification of home pages on the web. *Proceedings of the 38th Hawaii International Conference on System Sciences.* Los Alamitos IEEE Press.

Kress, Gunther / van Leeuwen, Theo 1996. *Reading Images. The Grammar of Visual Design.* London: Routledge.

Lemke, Jay L. 2002a. Travels in hypermodality. *Visual Communication*. 1/3, 299-325.

Lemke, Jay L. 2002b. Travels in hypermodality: Appendix. Distinctions and terms. Available at <http://www-personal.umich.edu/~jaylemke/papers/hypermodality/travels-App.htm>.

Lemke, Jay L. 2003. Texts and Discourses in the Technologies of Social Organization. In Weiss, Gilbert / Wodak, Ruth (eds), *Criti-*

cal Discourse Analysis. Theory and Interdisciplinarity. London: Palgrave/Macmillan, 130-149.

Lemke, Jay L. 2005. Multimedia genres and traversals. *Folia Linguistica* 39/1-2, 45-56.

Lemke, Jay L. 2009. Discursive Technologies and the Social Organization of Meaning. In Wodak, Ruth (ed.), Special Issue *Folia Linguistica*, Critical Discourse Analysis in Postmodern Societies. 35/1–2, 79-96.

Lindemann, Christoph / Littig, Lars 2011. Classification of Web Sites at Super-genre Level. In Mehler, Alexander / Sharoff, Serge / Santini, Marina (eds) *Genres on the Web. Computational Models and Empirical Studies*. Berlin: Springer Verlag, 211-236.

Linell, Per 1998. Discourse across boundaries: on recontextualizations and the blending of voices in professional discourse. *Text*, 18/2, 143-157.

Mehler, Alexander / Gleim, Rüdiger / Wegner, Armin 2007. Structural Uncertainty of Hypertext Types. An Empirical Study. In Rehm, Georg / Santini, Marina (eds) *Proceedings of the International Workshop Towards Genre-Enabled Search Engines: The Impact of Natural Language Processing*. Borovets, Bulgaria: INCOMA, 13-19.

Mehler, Alexander / Sharoff, Serge / Santini, Marina (eds) 2010. *Genres on the Web. Computational Models and Empirical Studies*. Berlin: Springer Verlag.

Orlikowski, Wanda J. / Yates, Jane 1994. Genre repertoire: The structuring of communication practices in organizations. *Administrative Science Quarterly* 39/4, 541-574.

Rauen, Fábio José 2009. Relevance and genre. Theoretical and conceptual interfaces. In Bazerman, Charles / Bonini, Adair / Figueiredo, Débora (eds) *Genre in a Changing World*. Perspectives on Writing. Fort Collins, Colorado: The WAC Clearinghouse and Parlor Press, 56-76.

Santini, Marina 2004. Identification of Genres on the Web: a Multi-Faceted Approach. *Proceedings of the ECIR 2004* (*26th European Conference on IR Research*), Volume 2, Poster Abstracts, Edited by Michael P. Oakes, University of Sunderland (UK), 5-7 April, 2004.

Santini, Marina 2005. Linguistic Facets for Genre and Text Type Identification: A Description of Linguistically-Motivated Features. *Technical Report ITRI-05-02*, 2005, ITRI, University of Brighton (UK).

Santini, Marina / Mehler, Alexander / Sharoff, Serge 2010. Riding the rough waves of genre on the web: Concepts and research questions. In Mehler, Alexander / Sharoff, Serge / Santini, Marina (eds) *Genres on the Web. Computational Models and Empirical Studies*. Dordrecht / Heidelberg / London / New York: Springer Verlag, 3-30.

Shepherd, Michael / Watters, Carolyn 1998. The evolution of cybergenres. *Proceedings of the 31st Hawaii International Conference on System Sciences*, Maui, Hawaii: IEEE Press, 97-109.

Shepherd, Michael / Watters, Carolyn / Kennedy, Alistair 2004. Cybergenre: automatic identification of homepages on the Web. *Journal of Web Engineering*. 3/3-4, 236-251.

Snyder, Ilana 1997. Beyond the hype: Reassessing hypertext. In Snyder, Ilana (ed.) *Page to screen: Taking literacy into the electronic era*. St. Leonards, NSW: Allen and Unwin, 125-143.

Spinuzzi, Clay 2003. *Tracing genres through organizations: A sociocultural approach to information design*. Cambridge, MA: MIT Press.

Spinuzzi, Clay 2004. Describing assemblages: Genre sets, systems, repertoires and ecologies. Computer Writing and Research Lab White Paper Series #040505-2, available at < http://www.dwrl.utexas.edu/files/040505-2.pdf>.

Spinuzzi, Clay / Zachry, Mark 2000. Genre ecologies: An open system approach to understanding and constructing documentation. *ACM J. Comput. Doc.* 24/3, 169-181.

Swales, John 1996. Occluded genres in the academy. In Ventola, Eija / Mauranen, Anna (eds) *Academic writing*. Amsterdam: John Benjamins, 45-58.

Swales, John 2004. *Research Genres. Explorations and Applications*. Cambridge: Cambridge University Press.

Symonenko, Svetlana 2007a. *Websites through genre lenses: Recognizing emergent regularities in websites content structure.*

iSchool Information Science and Technology – Dissertations and Theses. Paper 15. University of Syracuse.

Symonenko, Svetlana 2007b. Recognizing Genre-Like Regularities in Website Content Structure. In Rehm, Georg / Santini, Marina (eds), *Proceedings of the International Workshop Towards Genre Enabled Search Engines: The Impact of NLP*. Borovets, Bulgaria: Lecture Notes in Computer Science. 2007(4830) Short Papers, 29-34.

Watters, Carolyn / Dong, Lei 2006. Genre and Web search effectiveness. *Information Studies*. 12/3, 169-178.

Watters, Carolyn / Shepherd, Michael 1999. Cybergenre and Web functionality. *Proceedings of the 32nd annual Hawaii International conference on System Sciences*. Maui, Hawaii: IEEE Press.

Wodak, Ruth. 2006. Images in/and news in a globalised world. Introductory Thoughts. In Lassen, Inger / Strunck, Jeanne / Vestergaard, Torben (eds), *Mediating Ideology in Text and Image*, 1-16.

Yates, Jane / Orlikowski, Wanda 1992. Genres of organizational communication: A structurational approach to studying communication and media. *Academy of Management Review* 17/2, 299-326.

Yates, Jane / Orlikowski, Wanda J. / Renneker, J. 1997. Collaborative Genres for Collaboration: Genre Systems in Digital Media. *Proceedings of the 30th Annual Hawaii International Conference on System Sciences: Digital Documents*. Vol VI. Los Alamitos CA: IEEE Computer Society Press, 50-59.

Yates, Jane / Orlikowski, Wanda J. / Okamura, K. 1999. Explicit and Implicit Structuring of Genres in Electronic Communication: Reinforcement and Change of Social Interaction. *Organization Science*. 10/1, 83-103.

Yates, Jane / Orlikowski, Wanda 2002. Genre systems: Structuring interaction through communicative norms. *Journal of Business Communication*. 39/1, 13-35.

ALESSANDRA VICENTINI

Institutional Healthcare E-Brochures and Multilingualism Issues in the Recent Immigration Era in Italy (2007-2010)[1]

1. Introduction

Over the past three decades, the availability of new tools and technological approaches, together with the phenomenon of globalisation have rapidly and deeply altered the forms, contents, protagonists and role of (medical) communication (e.g. Mooney/Sarangi 2003; Hesse *et al.* 2005; Kress/van Leeuwen 2006; Fox/Jones 2009). Information flows have been so accelerated as to transform the models and modes of dissemination for medical and healthcare information, once only used exclusively or mostly by experts (Slevin 2002), and now available to the masses.

Web-mediated communication, in particular, has qualified as a powerful strategic resource in healthcare settings. It is central to any healthcare institution's effort to reach out to all possible users and patients pervasively and, though a large sector of the worldwide population, especially the elderly (Internet World Stats 2010),[2] does not

---

1 Part of the research carried out to lay down the theoretical framework of this chapter was conducted together with Kim Grego. This was done within a broader research project aiming to analyse how both Italian and international Anglophone healthcare websites reach out to users/citizens/potential patients with regard to their communicative and linguistic strategies (see Grego/Vicentini 2009, 2011, forthcoming; Vicentini *et al.* 2012).

2 According to Internet World Stats (2010), there were an estimated 1,966,514,816 Internet users in June 2010 (representing about 28.7% of the worldwide population), which is an increase of 444.8% compared to 2000, with Africa, Latin America and the Middle East taking the lion's share.

access the Internet yet, it has become a fact nowadays. Through the web, users/potential patients are now provided with all types of information and a full range of possible services from which to choose, with convenient times, treatments and costs; healthcare/medical products and services (e.g. medications, drugs, diagnostic and genetic tests, etc.) are now directly available via the Internet, while beforehand their access and practice were mediated by healthcare operators (Orizio *et al.* 2010a). All this, besides creating a new type of self-informed patient posing ethical dilemmas already referred to as *e-health ethics* (Eysenbach 2001; Andreassen *et al.* 2007), has significantly modified the marketing and communication strategies deployed by both healthcare suppliers and institutional healthcare systems, which have been gradually overlapping with the business sector (cf. Baru 2003: 4433; May/ Cheney/Roper 2007; Orizio *et al.* 2010b). Indeed, these have in common a clear customer orientation, with the identification of potential clients that are likely to be impressed by a display of efficiency, readiness, and service availability as well as trustworthiness, ethical behaviour, and a sustainable approach (Vicentini 2010).

Furthermore, the aforementioned evolution has given rise to phenomena of genre and language change, which involve all professional settings at large, and the healthcare sphere in particular (suffice it to think of relatively recent, yet well established, terms such as *e-health*[3] and *medicine 2.0*, which aptly represent the new communicative paradigm illustrated above). It is well known, by now, that genres have been considerably affected by the Internet turn, which has added new media and modes to the standard, conventional 'print and pictures' dimension as well as widening and diversifying participation frameworks as never before. Scholars generally agree on the fact that genre migration to the web together with the rise of new web-specific genres have contributed to extending and modifying the concept of

3 The term, testifying to the transformation of medicine in the Internet era, was defined in 2001 by the editor of the *Journal of Medical Internet Research* (the leading journal in the field) as follows: "e-health is an emerging field in the intersection of medical informatics, public health and business, referring to health services and information delivered or enhanced through the Internet and related technologies" (Eysenbach 2001: 20).

genre to make it suitable for the new digital media (cf. Garzone 2007 and references therein).

A number of labels have been used so far to describe the transitory status of genres after these have joined the online constituent. To name but a few: *genre hybridisation, remediation, repurposing, shift, bending, chains, mixing, juxtaposition, multigenre*, etc. In online institutional and corporate healthcare communication, especially, hybridisation as a trend is increasing through the use of hybridised genres and discourses. Indeed, there have emerged some discursive practices originally belonging to other specialised discourses – business and tourism specifically (Catenaccio 2009; Vicentini 2010) – and their traditionally associated genres (e.g. brochure, corporate press release, organisation chart, etc.), a phenomenon that highlights a variation in purpose too.

Moreover, global forces such as the presence of emerging market countries and the ensuing transnational circulation of populations, cultures, ideas, and languages (Held/McGrew 2003; Maurais/Morris 2003; Risager 2006) are progressively reshaping (healthcare) communication strategies and language, while struggling against local forces and autonomies.

Italy's public healthcare, constantly disputed between central and local government, is a case in point in this respect. Due to the recent (2007-2010)[4] phenomenon of immigration to the country, on top of the new trends described above, it has undergone considerable changes in the way it communicates via the web, having to tackle new and complex issues of multilingualism and cultural mediation. This fact can reveal new insights into the complicated relation between identity, language, and communication and is thus worth analysing here.

2. Background and aim

The recent flows of immigrants to Italy have dramatically contributed to increasing the country's already fragmented linguistic situation

4 The material used for this study was last accessed on 7/01/2010.

resulting from (a) the presence of traditionally bi-/multilingual regions, (b) the hundreds of diatopic varieties (dialects) of Italian not enjoying language status (Berruto 1987: 27) and (c) the use of international communication languages (i.e. EU working languages) in professional settings. Official statistics (ISTAT 2008, 2009), in particular, report a significant increase in the presence of immigrants from EU countries like Romania and Bulgaria[5] starting from 2007, though non-EU immigrants from both within (i.e. Albania, Ukraine and Moldova) and outside (i.e. Morocco, China, the Philippines, Sri Lanka, Brazil and Peru) Europe had outnumbered them by 2009, thus generating an even more complex and multifaceted linguistic scenario.[6]

Immigration phenomena usually give rise to demographic behaviours and trends – like getting married and bearing children – that are typical of a settled population as well as a number of new necessities to be confronted by policy makers. Among these, the access to public healthcare structures can reveal interesting data concerning issues of stability and change vis-à-vis immigrants. In Italy, statistics reveal that foreigners are increasingly turning to public structures to be treated: men have been seen to address first aid services in particular, while women seek medical assistance at consulting rooms.

In light of the above, this chapter will scrutinise all of Italy's regional healthcare websites from the perspective of a highly relevant identity-defining descriptor, that is to say multilingualism. This is an element that can contribute to gaining some information on the way public health institutions have tackled and still tackle the recent flows of immigrants from Eastern European, Far Eastern, North African and Arab countries to Italy.

Through the investigation of this trend – albeit in a short-term perspective (2007-2010) – the final intent is to show what kind of informative material is at the immigrants' disposal, and if/how it has changed over time in terms of purpose, form and contents. In particular, the new web-based institutional healthcare (e)brochure genre will

---

5 These countries officially joined the European Union on 1/01/2007.

6 Only legal immigrants are considered by official statistics. Illegal ones are not even mentioned in official documents: the real socio-linguistic situation, therefore, must be far more complicated.

be considered, describing its genre, textual, communicative, and linguistic features when targeted at a specific social/ethnic group (i.e. the recent immigrants to Italy).

2.1. The institutional healthcare brochure genre

Hitherto no studies have been carried out on the institutional healthcare/medical (e)brochure genre, and only a few deal with its paper counterpart, which the online version clearly derives from and is related to. Research has concentrated on the brochure genre in the business sector, as part of corporate communication. Askehave (1998) and Askehave/Swales (2001), working with the tools of systemic functional analysis and genre analysis, have shown how company brochures, once used to depict products produced or services offered, now aim to establish credentials and present the company as a qualified partner to prospects and suppliers. Shepherd/Watters (1999) briefly describe cybergenres as a new type of genres characterised by three features: content, form, and functionality. Among these, the web-based company brochure is included. This is compared to homepages with promotional/advertising contents. The authors argue that this new cybergenre is very similar to the single-page, tri-folded paper brochure, which had, right at the outset of the migration of traditional genres to the web, a quite flat and linear structure, with much visual impact and only marginal functionality, i.e. browsing and e-mailing. Finally, Grego (2010) takes into account an example of the corporate medical (e)brochure genre, demonstrating how the shift towards the popularisation of science has affected the field of healthcare and medicine too, especially as communicated from corporation to citizen on the web. Grego describes how this is carried out through heavy forms of (purpose, discourse and genre) hybridisation.

The void in research into the electronic format of the institutional medical/healthcare brochure leads us to take into account its paper equivalent as our reference model for the present analysis. Institutional healthcare/medical brochures[7] have so far been analysed by focussing on their communicative purposes – informative or educational or both – as well as on their discursive and linguistic features. They represent a convenient information source for users just starting their information search on a specific medical issue. They are generally authored by national healthcare institutions/organisations or social /educational bodies and targeted at all the potential users/patients of the latter's reference discourse community, whether national or local. They are usually structured as tri-fold brochures, consisting of a front cover, an inside front cover, inside page(s) and back page(s), enacting a prototypical move structure typically organised as follows:

1. identifying sender;
2. attracting attention of the reader to the topic;
3. detailing contents;
4. making recommendations to users/potential patients;
5. establishing contacts.

Since such material is meant to help users find the information needed with ease, the above moves are textually organised in paragraphs, bulleted areas, titles and subtitles. Visual contents are more often than not employed to demonstrate a specific area of the body, show symptoms and enhance the ability to understand the issues. Charts, graphs and other visuals can be used to diagram and show medical statistics depending on the topic, but usually balance, and never overwhelm, the writing parts. The above sketched genre model will be used as a benchmark for the analysis in the next paragraphs.

7 A research project on informative medical brochures has been conducted and is still under way at the Writing in Digital Environments (WIDE) Research Center of the Michigan State University under the supervision of Prof. Hart-Davidson (see Heinze 2004).

3. Corpus and method

The corpus consists of 178 websites of ASL (*Aziende sanitarie locali*, literally: Local Health Enterprises), or Healthcare District Units (HDUs), and 95 websites of AO (*Aziende ospedaliere*, literally: Hospital Enterprises), or Hospital Administrations (HAs), both groups making up the operative part of the INHS.[8] The starting point for website selection was the site of the Ministry of Health,[9] which features links to both these national and local organisations.[10] Worthy of note is also the fact that they are all public entities, 97% of which are publicly funded, and that only official websites have been considered. The corpus was first collected and analysed in 2007; a second analysis was carried out in 2010 so as to verify the evolution of the informative material directed at immigrants in terms of quantity and quality.

The investigation took time, since the websites were first scrutinised manually to identify the specific material/information directed at foreigners. The results were then interpreted in terms of move structure, purpose, contents, form (layout, textual organisation) and language, working from the perspective of critical discourse analysis (Fairclough 1995, 2003), genre analysis (Swales 1990; Bhatia 1993) and multimodality (Kress/van Leeuwen 2006; Garzone 2007). Additional insights were gained from sociolinguistic tools (Berruto 1987; Meyerhoff 2006) and from specific literature on web genres (cf. § 1).

---

8 Abbreviation for Italian National Healthcare System (*Servizio Sanitario Nazionale*).

9 The Italian Ministry of Health (*Ministero della Salute*), <http://www.ministerosalute.it>.

10 Their number (178 HDUs and 95 HAs corresponding to as many websites) did not change over the time span considered.

4. Analysis and discussion

Contrary to the expectations justified earlier (§ 2), the analysis reveals that no HA and only a few HDU websites offer information in the recent immigrants' languages in the period under scrutiny. In particular, in 2007 only 15 (i.e. 8.4 %) out of 178 HDU websites displayed information in the recent immigration languages. These all belonged – apart from Napoli 1 HDU, Campania – to the HDUs of the Northern-Italian territories (i.e. Valle d'Aosta, Lombardia, Friuli Venezia Giulia, Veneto and Emilia Romagna), where the influx of immigrants in search of a job has mainly converged in the last years. Veneto, a region with a significant presence of immigrants,[11] unexpectedly featured only one HDU, while Emilia Romagna took the lion's share, since 6 out of its 11 HDUs provided healthcare communication to immigrants. This is in line with a trend the latter region has been and is still experiencing: foreign immigration is in fact transforming its demographic and social face quite rapidly.[12] Finally, the presence of Friuli Venezia Giulia should not surprise, as this Region is familiar by nature with multiethnicity and multilingualism issues. Figures show a slight increase to 24 (i.e. 13.5%) websites in 2010, with the addition of the Province of Trento, Marche, and some more Veneto HDU websites providing information in the new immigrants' languages. Considering that almost 4.3 million immigrants are officially living in Italy (ISTAT 2009), which represents 7% of the entire Italian popula-

11 In Veneto and Lombardy one worker out of two is an immigrant, in particular immigrants represent 35% of the entire working population in the building industry in Veneto (Stranieri in Italia, <http://www.stranieriinitalia.it/news/ statistiche.html>). According to ISTAT data, in 2008 legal immigrants ranked as follows: 521,295 in Lombardy, 250,734 in Emilia-Romagna, 243,886 in Veneto, 200,574 in Lazio, 174,217 in Tuscany, 149,571 in Piedmont. The first Southern region was Campania with 84,336 immigrants, who are definitely remarkably fewer in the South (Permessi di soggiorno, ISTAT, <http://demo.istat.it/altridati/permessi/index.html>).

12 Emilia Romagna's population is expected to increase by 1 million people in the next 20 years, thus growing ten times faster than in the previous century (Bonaguidi/Valentini 2007).

tion, the situation described above definitely indicates a lack, or at least a scarcity, of provisions to develop suitable healthcare communication strategies for foreigners, despite slight evidence of progress.

Table 1 below reports detailed data regarding the specific language(s) used in the sections hosted by the HDU websites examined as well as the exact part/section of the website where they are located. It shows, in particular, that neither whole nor partial website versions in any of the recent immigrants' languages are provided, but only sections, that is to say links to pdf or Word documents/files to be downloaded, which amounted to 96 occurrences in 2007, with a significant 30% increase to 140 in 2010. This means that information in the recent immigrants' languages is never displayed in prominent (level 1 or 2) webpages/locations, which are directly accessible from the homepage. In fact, this information can only be reached by surfing the deeper layers of each website, carefully looking for the desired material, thus proving that reaching out to immigrants is not the Italian healthcare institutes' priority.

Table 1 shows that the most frequent recent immigration languages are Arabic, Chinese, Russian, Albanian, Romanian and Spanish. These data do not change in the period taken into account, and certainly echo official statistics on immigration to Italy in recent times (see § 2). In particular, under the label 'Arabic', diverse Middle East, Central-Asian and North African countries are to be included, while 'Spanish' also covers Central-Southern America. Though most Romanians speak Italian, their immigration has so deeply affected Italian society that their linguistic presence clearly emerges from the public webpages that are meant to be addressed to foreigners. Statistical data indicating how massive the immigration from Central-Eastern Europe[13] has been are confirmed by the occurrence of documents in Russian, Albanian and Ukrainian.

13 See 'Permessi di soggiorno dei cittadini extracomunitari per sesso, area geografica e singolo paese di cittadinanza al 1° gennaio 2008', Permessi di soggiorno, ISTAT 2008, <http://demo.istat.it/altridati/permessi/>, for data concerning the recent immigrants' countries of origin.

Language[14]	*2007* *No. of Sections*[15]	*2010* *No. of Sections*	*Language*	*2007* *No. of Sections*	*2010* *No. of Sections*
Albanian	11	15	Russian	12	13
Amharic	1	3	Serbo-Croatian	2	3
Arabic	15	19	Slovenian	1	2
Bengali	1	2	Somali	2	2
Chinese	13	20	Spanish	9	13
Croatian	1	3	Sri Lankan	1	1
Gujarati	1	4	Swahili	1	1
Hindi	1	5	Tamil	1	3
Macedon	2	4	Tigrinya	1	3
Mirpuri	1	3	Turkish	2	4
Polish	3	5	Ukrainian	2	4
Portuguese	5	5	Urdu	2 + 1[16]	2 + 1
Punjabi	1	2	Vietnamese	1	1
Romanian	9	9			

Table 1. Recent immigration languages as found in the Italian HDU websites listed in alphabetical order – sections of websites (2007-2010).

Apart from some general data providing guides to hospitals (i.e. structures, facilities, organizations, etc.), most multilingual documents give practical indications often directed at different ethnic groups (thus provided in as many different language versions), because of either dissimilar culture habits or lifestyle/working environment, which might impact on the immigrants' health conditions. Indeed, specific advice addressed to (home) carers is mostly provided in Eastern-European languages (especially Ukrainian, Russian and Polish); healthcare suggestions on how to behave with infants and new-born babies are given in Spanish, Portuguese, Punjabi, Urdu and Mirpuri,

14 The total number of HDU websites considered is 173. In actual fact the total number of websites existing at the time of scrutiny was 178, but five websites could not be accessed for technical reasons (i.e. connection problems, web re-styling and reorganization, etc.).

15 The format of the sections varies: pdf, Word, html links, etc.

16 These Urdu speakers were labelled as 'Pakistani'.

most probably responding to the needs of Central-Southern American and Asian mothers/families, who are significantly contributing to the latest Italian demographic increase; information concerning female genital mutilation is made available in Somali, Amharic, Swahili, Tigrinya and Arabic. Nevertheless, all HDU html webpages directed at immigrants are only in Italian and, even when multilingual documents are available, indications on the link and web-path to follow in order to reach the file searched for are again given in Italian only. The reasons for this are multiple: firstly, multilingual documents, being usually provided by outsourced translators, are more easily uploaded/attached in pdf/Word formats than as html webpages; as such, they can also be more easily changed, thus better responding to the administrators' needs; finally, since immigrants are still thought of as not having access to IT facilities and Internet connections, they still depend on a hypothetical Italian employer/friend, who can help them download specific healthcare documents, which are in fact still conceived as paper leaflets, easy to download, to be passed on to the immigrant and consulted.

As regards the specific pdf/Word material for foreigners, it mainly consists of multilingual informative brochures and leaflets totalling 162 elements, which tend to increase over the years.

4.1. The institutional healthcare (e)brochure genre for foreigners

In order to classify and analyse the above corpus, so as to establish the genre features of the institutional healthcare (e)brochure for foreigners (i.e. the new immigrants to Italy), some generic features were selected and then interpreted (see Table 2). The language versions provided by the websites were considered to assess possible changes in the number and typology of addressees/target users (i.e. audience); the texts' move structure and purposes were also analysed to see whether their social function had changed over time; then also some structural and linguistic features were looked at, that is grammar (i.e. short/long, minor/major sentences, simplified syntax, finite/non-finite verb forms, etc.) and lexical elements (i.e. rare, learned, technical lexicon vs. simple, everyday terms, etc.), to verify the type of linguistic strategies and register deployed.

The results point to an unchanged target user, i.e. the number and typology of the new immigrants have not varied. The same can be said for purpose, which remains mainly educational/informative. There are no persuasive elements, typical of marketing strategies, which means that immigrants are not seen as target clients to attract or persuade. This differs from the general marketing-oriented trend that also publicly-funded healthcare institutions have been undergoing in the last few years (see § 1). Indeed, the move structure enacted by such diverse brochures has not varied through the two years considered as compared to the paper version taken as a prototypical reference model in § 2.1. It is thus evident that what these kinds of texts try to accomplish within their specific social context has not varied, i.e. they inform the target user either on the services provided by the healthcare public institution or on some specific diseases/conditions. Worthy of note, however, is that move 3 (i.e. detailing contents, Table 2) has been reduced significantly: contents were more thoroughly detailed in the 2007 corpus as well as in the paper, prototypical model. This means that, when (e)brochures are directed at the new immigrants, information density tends to decrease and, consequently, such documents do not act as explanatory texts the user can rely on to learn about a specific disease/condition anymore. They just prove to be a first-contact tool to get in touch with the reference healthcare institution, an element that reinforces move 5 (establishing contact, Table 2).

INDICATORS	**2007**	**2010**
AUDIENCE		
No. of language versions	27	27
MOVES		
	1. Identifying sender	1. Identifying sender
	2. Attracting attention of the reader to the topic	2. Attracting attention of the reader to the topic
	3. Detailing contents	3. Detailing contents
	4. Making recommendations to users/potential patients	4. Making recommendations to users/potential patients
	5. Establishing contacts	**5. Establishing contacts**
PURPOSE		
Informative/educational	100%	100%
Persuasive/promotional	0%	0%
STRUCTURE AND TEXTUAL FEATURES		
Format: pdf	95.55%	99.28%
Format: Word	4.45%	0.72%
Images/visual contents	40%	70%
two or more languages in the same document	4.44%	8.57%
REGISTER		
Informal	50%	80%
Formal	50%	20%

Table 2. Generic features of the e-brochures directed at recent immigrants to Italy (2007-2010).

Turning to structure, the pdf – a frozen text, without hyperlinks, which users cannot interact with – proves to be the specific format/channel to communicate via the web with the new immigrants in the Italian healthcare environment.[17] Visual contents (i.e. graphics, images, pictures, colours, diagrams, etc.) are given more and more space: not only have they increased by 30%, but they have also qualitatively changed. In fact, images tend to be more stylised, they do not refer to any specific ethnic groups, either for political correctness or for economic concerns, as the same graphic design and layout can be used for different languages, thus saving time and money. Their function has also changed: visuals do not act as complementary elements to the written parts of the text, i.e. they do not just illustrate what the written sections express, but they have become the main semiotic code through which the message is conveyed. Such a development is in line with a gradual simplification in the language used, i.e. simple or even minor sentences, contractions, enumeration, non-finite verb-forms, etc. Lastly, while in the 2007 reference corpus there are as many brochures as the languages provided, in 2010 preference is given to one single multilingual brochure, which is again in keeping with the general tendency toward simplification.

A few examples representative of the 2007 and 2010 sets of brochures respectively will suffice to illustrate the data presented and discussed above. Figure 1 illustrates the typical informative medical (e)brochure featured in the 2007 set of documents. It is a two-page brochure used to inform patients about the GP's and the paediatrician's activities. It is offered in five different versions: English, French, Chinese, Arabic and Italian. The written code is prevalent, the text is organised in columns with many bulleted areas to help readers find information more easily. At the top of the page, the logo, which usually accounts for the authorship or sponsorship of the brochure (move 1, Table 2), is in an English-Italian version. Contents (move 3, Table 2) are quite detailed and dense. Visuals embody real people framing the written code. It is possible that they have been inserted to reinforce the text's main topic, as well as reassure readers, representing as they do the institutional/official element. The same layout and graphic vs. writing structure was used in all the five different language versions provided.

17 Brochures in Word format were scarce, and thus not considered significant.

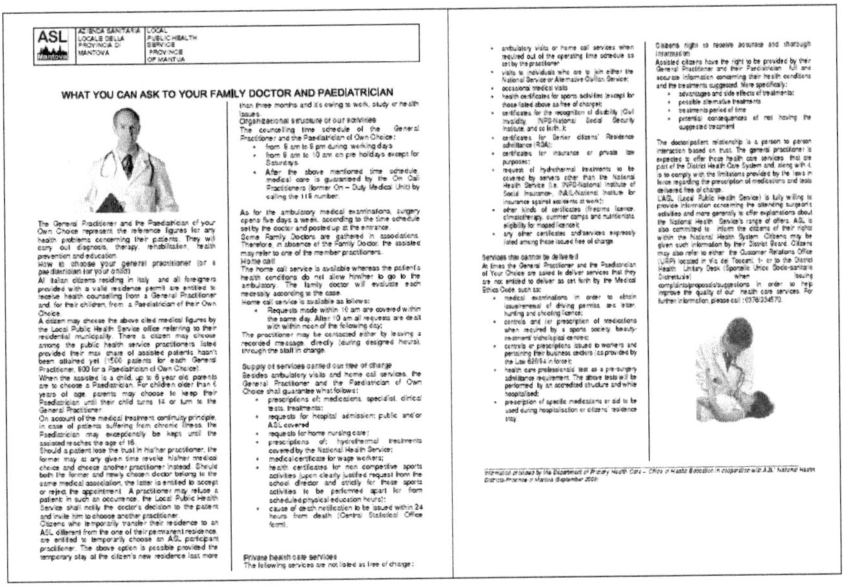

ASL Mantova | AZIENDA SANITARIA LOCALE DELLA PROVINCIA DI MANTOVA | LOCAL PUBLIC HEALTH SERVICE PROVINCE OF MANTUA

WHAT YOU CAN ASK TO YOUR FAMILY DOCTOR AND PAEDIATRICIAN

The General Practitioner and the Paediatrician of your Own Choice represent the reference figures for any health problems concerning their patients. They will carry out diagnosis, therapy, rehabilitation, health prevention and education.

How to choose your general practitioner (or a paediatrician for your child)

All Italian citizens residing in Italy and all foreigners provided with a valid residence permit are entitled to receive health counselling from a General Practitioner and, for their children, from a Paediatrician of their Own Choice.

A citizen may choose the above cited medical figures by the Local Public Health Service office referring to their residential municipality. There a citizen may choose among the public health service practitioners listed provided their max share of assisted patients hasn't been attained yet (1500 patients for each General Practitioner, 800 for a Paediatrician of Own Choice).

When the assisted is a child, up to 6 year old, parents are to choose a Paediatrician. For children older than 6 years of age, parents may choose to keep their Paediatrician until their child turns 14 or turn to the General Practitioner.

On account of the medical treatment continuity principle, in case of patients suffering from chronic illness, the Paediatrician may exceptionally be kept until the assisted reaches the age of 16.

Should a patient lose the trust in his/her practitioner, the former may at any given time revoke his/her medical choice and choose another practitioner instead. Should both the former and newly chosen doctor belong to the same medical association, the latter is entitled to accept or reject the appointment. A practitioner may refuse a patient: in such an occurrence, the Local Public Health Service shall notify the doctor's decision to the patient and invite him to choose another practitioner.

Citizens who temporarily transfer their residence to an ASL different from the one of their permanent residence, are entitled to temporarily choose an ASL participant practitioner. The above option is possible provided the temporary stay at the citizen's new residence last more than three months and it's owing to work, study or health issues.

Organizational structure of our activities

The counselling time schedule of the General Practitioner and the Paediatrician of Own Choice:

- from 8 am to 8 pm during working days
- from 8 am to 10 am on pre holidays except for Saturdays.
- After the above mentioned time schedule, medical care is guaranteed by the On Call Practitioners (former On – Duty Medical Unit) by calling the 118 number.

As for the ambulatory medical examinations, surgery opens five days a week, according to the time schedule set by the doctor and posted up at the entrance.

Some Family Doctors are gathered in associations. Therefore, in absence of the Family Doctor, the assisted may refer to one of the member practitioners.

Home call

The home call service is available whereas the patient's health conditions do not allow him/her to go to the ambulatory. The family doctor will evaluate each necessity according to the case.

Home call service is available as follows:

- Requests made within 10 am are covered within the same day. After 10 am all requests are dealt with within noon of the following day;

The practitioner may be contacted either by leaving a recorded message, directly (during designed hours), through the staff in charge.

Supply of services carried out free of charge

Besides ambulatory visits and home call services, the General Practitioner and the Paediatrician of Own Choice shall guarantee what follows:

- prescriptions of: medications, specialist, clinical tests, treatments;
- requests for hospital admission: public and/or ASL covered
- requests for home nursing care;
- prescriptions of: hydrothermal treatments covered by the National Health Service;
- medical certificate for wage workers;
- health certificates for non competitive sports activities (upon clearly justified request from the school director and strictly for those sports activities to be performed apart for from scheduled physical education hours);
- cause of death notification to be issued within 24 hours from death (Central Statistical Office form).

Private health care services

The following services are not listed as free of charge:

- ambulatory visits or home call services when required out of the operating time schedule as set by the practitioner
- visits to individuals who are to join either the National Service or Alternative Civilian Service;
- occasional medical visits
- health certificates for sports activities (except for those listed above as free of charge);
- certificates for the recognition of disability (Civil invalidity, INPS-National Social Security Institute, and so forth...);
- certificates for Senior citizens' Residence admittance (RSA);
- certificates for insurance or private law purposes;
- request of hydrothermal treatments to be covered by servers other than the National Health Service (i.e. INPS-National Institute of Social Insurance-, INAIL-National Institute for insurance against accidents at work);
- other kinds of certificates (firearms licence, climatotherapy, summer camps and nutritionists eligibility for moped licence);
- any other certificates and/services expressly listed among those issued free of charge.

Services that cannot be delivered

At times the General Practitioner and the Paediatrician of Your Choice are asked to deliver services that they are not entitled to deliver as set forth by the Medical Ethics Code, such as:

- medical examinations in order to obtain issue/renewal of driving permits, arms letter, hunting and shooting licence;
- controls and /or prescription of medications when required by a sports society beauty-treatment/ trichological centres;
- controls or prescriptions issued to workers and pertaining their business sectors (as provided by the Law 626/94 in force);
- health care professionals' test as a pre-surgery admittance requirement. The above tests will be performed by an accredited structure and while hospitalised;
- prescription of specific medications or aid to be used during hospitalisation or citizens' residence stay.

Citizens right to receive accurate and thorough information

Assisted citizens have the right to be provided by their General Practitioner and their Paediatrician full and accurate information concerning their health conditions and the treatments suggested. More specifically:

- advantages and side effects of treatments;
- possible alternative treatments
- treatments period of time
- potential consequences of not having the suggested treatment

The doctor/patient relationship is a person to person interaction based on trust. The general practitioner is expected to offer those health care services that are part of the District Health Care System and, along with it, is to comply with the limitations provided by the laws in force regarding the prescription of medications and tests delivered free of charge.

L'ASL (Local Public Health Service) is fully willing to provide information concerning the attending surgeon's activities and more generally to offer explanations about the National Health Service's range of offers. ASL is also committed to inform the citizens of their rights within the National Health System. Citizens may be given such information by their District Board. Citizens may also refer to either the Customer Relations Office (URP) located in Via dei Toscani, 1- or to the District Health Unitary Desk (Sportello Unico Socio-sanitario Distrettuale) when issuing complaints/proposals/suggestions in order to help improve the quality of our health care services. For further information, please call :0376/334570.

Information provided by the Department of Primary Health Care – Office of Health Education in cooperation with ASL' National Health Districts-Province of Mantua (September 2006)

Figure 1. Example of a 2007 brochure– 'What you can ask to [sic] your family doctor and paediatrician', Mantova HDU, Lombardia, English version (7/06/2007).

Some of these considerations apply to Figure 2 too, which illustrates the brochure given by Bergamo HDU to explain what the family voucher is and how to obtain it. This time images consist of tables and grids containing specific economic and contact indications. Information is organised through FAQs, proving to be direct and clear.

My child will be 7 in 2009, may I require the voucher?
No, if a child will be 7 in the year 2009, the family is not entitled to require the voucher

I have three children, one is 9, one is 6 and one will be 18 in 2009. May I require the voucher?
No, if one of the children will be 18 in the year 2009 the family is not entitled to require the voucher.

WHERE WILL I HAND IN APPLICATIONS?
Applications may be handed in:

- at ASL local offices;

WHEN MAY I HAND IN APPLICATIONS?
Applications may be handed in from February 16 to March 13, 2009

HOW WILL I KNOW WETHER I HAVE RECEIVED THE VOUCHER?

From February 16 ASL will inform the eligible families that the grant has been directly paid on the Bank Account or by Postal Order.

10.000 (ten thousand) euro of ISR are the equivalent of the figures of this table

Family members	Two parents: only one parent working	Both parents working	One parent: working	One parent, not working: with a disabled person	One parent, not working and with a disabled person	Both parents working: with a disabled person
4	==	==	28.600	31.600	33.600	==
5	28.500	30.500	32.500	35.500	37.500	35.500
6	32.000	34.000	36.000	39.000	41.000	39.000
7	35.500	37.500	39.500	42.500	44.500	42.500
8	39.000	41.000	43.000	46.000	48.000	46.000
9	42.500	44.500	46.500	49.500	51.500	49.500
10	46.00	48.000	50.000	53.000	55.000	53.00

What the FAMILY VOUCHER is and how to obtain it from

RegioneLombardia

A **Family Voucher** is a grant from Regione Lombardia to low-income families with three or more children to support their cares.

WHO IS ELIGIBLE FOR IT?
The families with these features:

1) presence of at least three underage children, including those in family trust/fostering, one should be of at least 0-6 years of age.

2) families as per the Italian Constitution Chart (art. 29 "...based on marriage"), Regione Lombardia Statute and regional act 23/39 "Family policy"

3) the applicant parent and all of the children must live in Lombardia

The applicant parent must be:

- an Italian citizen or a European citizen registered at the local council;

- a foreign citizen with a EC long-time residence paper or permit

- The Income Stand Index must not be higher than 10.000 (ten thousand euro)
(Delibera di Giunta Regionale 20.01.2009 n. VIII/8881)

For information:

Call Center Regionale
840.00.00.01
Asl Bergamo
Ufficio Relazioni con il Pubblico
800.44.77.22
Azienda sanitaria Locale
Della Provincia di Bergamo
Via Galliccioli 4 24121 Bergamo

WHERE	WHEN
Bergamo Via Borgo Palazzo, 130 Padiglione Giallo 6D Tel. 035 2270325 035 2270334	From Monday to Friday 8.30-13 From Monday to Thursday 14-16
Dalmine Viale Betelli, 2 tel. 035 378105	From Monday to Friday 9-12
Zanica Via Serio, 1 tel. 035 4246581	From Monday to Friday 9-12
Ponte S. Pietro Via Caironi, 7 tel. 035 603329	From Monday to Friday 9-12
Villa D'Almè Via Roma, 16 tel. 035 634615/617/620	From Monday to Friday 9-12
Zogno VIA P. POLLI, 16 tel. 0345 59165/120	From Monday to Friday 9-12
Sant'Omobono Terme Via V. Veneto, 131/A tel. 035 851052	Tuesday 14-15,30
Piazza Brembana Via Monte Sole, 2 tel. 0345 81724	Tuesday 14-15,30
Treviglio Viale Partigiano, 5 Tel. 0363 304141	Da lunedì a venerdì 9-12 e 14-16
Romano di Lombardia Via Cavagnari, 5 0363/919219	Da lunedì a venerdì 9-12 e 14-16
tel. 035 3235036	10-12 + Monday and Thursday 14-16
Grumello del Monte Via Nembrini, 1 tel. 035 8356324 tel. 035 8356332	Monday, Thursday, Friday 10-12 + Monday e Thursday 14-16
Trescore Balneario Via Mazzini 13 tel. 035 955436/9	From Monday to Friday 8,30-12.30
Sarnico Viale Libertà, 37 tel. 035 911026	From Monday to Friday 8,30-12,30
Lovere Via Bonomelli, 5 tel. 035 4349611	From Monday to Friday 8,30-12,30 Mon. Tue. Thu. 13.30-16.00
Alzano Via Ribolla 3 035/4285252	From Monday to Friday 8.30-12,30 Mon. Tue. Thu. 14.00-15.30
Albino Viale Stazione 26/a 035/75.97.00	From Monday to Friday 8.30/12.30 Mon. Wed. Thu. 14.00/15.30
Gazzaniga Via Manzoni 98 035/71.41.42	Mon. Wed. Thu. 8.30/12.30
Gandino Via S.Bosco 2 035/74.62.53	Tuesday and Friday 8.30/12.30
Ponte Nossa Viale Partigiani 5 035/70.11.26	Friday 8.30/12.30
Clusone Via Matteotti 11 0346/89.029	From Monday to Friday 8.30/12.30 Mon. Wed. Thu 14.00/15.30
Gromo P.zza Pertini 1 0346/41.105	Wednesday 8.30/12.30
Vilminore Via Polini 7 0346/51019	Tuesday 9.00/12.30

Figure 2. Example of a 2007 brochure – 'What the family voucher is and how to obtain it', Bergamo HDU, Lombardia, English version (7/06/2007).

When moving on to the 2010 set of documents, just a glance is sufficient to highlight the differences from the 2007 corpus. Figure 3 represents the pdf brochure made available by Piacenza HDU, explaining how to assist the elderly. Visual communication is certainly prevalent, pictures are noticeably stylised; the written code is extremely simplified through the use of deverbal nouns (e.g. "moving"), imperatives (e.g. "bend his knees") and enumeration.

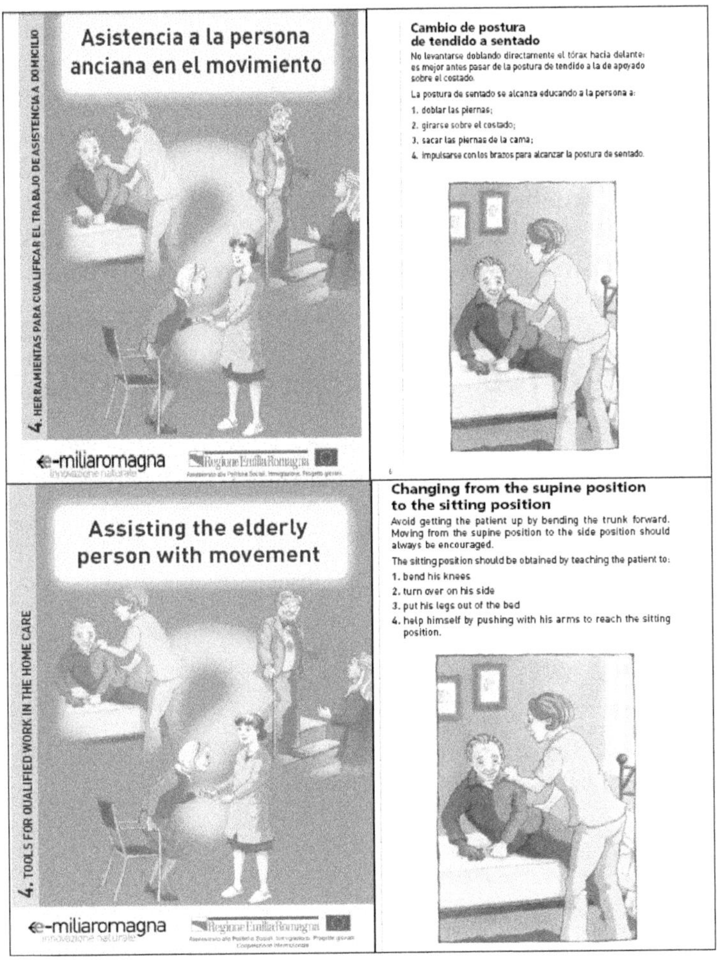

Figure 3. Example of a 2010 brochure – 'Assisting the elderly person with movement', Piacenza HDU, Emilia Romagna, Spanish and English versions (7/01/2010).

The brochure supplied by Forlì HDU giving information to foreign women on how to face their pregnancy (Figure 4) is even more stylised. Images have almost completely replaced words, which are used like slogans ("we're with you"), playing on different letterings and positions in the page (at the top and on the left-hand side), and thus behaving like visuals themselves, ultimately proving that the two codes (written and visual) are completely interchangeable here. This is confirmed by the fact that the stylised picture occupies the centrepiece position – the place where the most important information is usually presented (cf. Kress/van Leeuwen 2006) – and that it is a visual metaphor, meaning that the healthcare structure is supporting foreign pregnant women, a notion symbolised by a huge hand literally bearing the figure of a woman. Finally, also the way the written text is organised seems to embrace, and thus metaphorically support, the figure of the woman.

Figure 4. Example of a 2010 brochure – 'Facing your pregnancy', Forlì HDU, Emilia Romagna, English and Russian versions (7/01/2010).

5. Conclusions

The identity profile that emerges from the corpus reflects a deeply Italian-centred society as far as healthcare is concerned (see §1). The increase in the number of immigrants to Italy in the period considered does not coincide with a corresponding presence of healthcare information provided through the Internet in their languages. Proof of this is the relevant number of webpages containing specific information designed for foreign citizens – or even offering linguistic mediation services – written in Italian only, which makes one wonder how a foreigner with little or no command of Italian can ever access the online information that was provided to help him or her. Even when foreign language documents exist and are available on an HDU or HA website, very often the link to the downloadable document is paradoxically an Italian phrase, e.g. the name of a language not in the language itself but in Italian, thus preventing accessibility and usability.

However, though inconsistent, the presence of a variety of brochures and guides in a variety of extra-European idioms certainly shows a novel willingness to reach out to specific groups of healthcare users who are otherwise difficult, or impossible, to reach through Italian. It is still in an embryonic state, nevertheless, compared to the 12-language version offered by the UK National Health Service website (NHS, <www.nhs.uk>), for example. The quantitative research lying at the basis of this chapter, in particular, has highlighted a slight increase in the numbers of multilingual documents directed at recent immigrants to Italy in the period 2007-2010, which confirms that the phenomenon of immigration is definitely still in progress and that Italy is gradually becoming aware of it.

It emerges that there exists a specialised genre to inform and educate the new immigrants to Italy on healthcare issues, that is the pdf (e)brochure, here defined as institutional healthcare (e)brochure for foreigners (i.e. the new immigrants to Italy). Though many of the typical capabilities and functionalities of web-mediated genres (i.e. hypertextuality, browsing, e-mailing, etc.) are not as yet present, since these pdf texts are frozen documents that can be printed out and read/used as if they were paper versions, or read directly on screen, the investigation has nonetheless pointed out some significant changes/adaptations both as compared to the paper version taken as a prototypical model and in the brief period considered.

Simplification is the main tendency the diachronic analysis has pointed out. This is apparent at various language levels. Though the texts' general (informative/educational) purpose has remained unvaried throughout the years, and also as compared to its paper counterpart, the general reduction – in terms of presence and quantity of words – of move 3 (detailing contents, Table 2), reveals a slight overall functional shift. These (e)brochures seem to act just as a gateway to getting in touch with the healthcare institutions: far from being exhaustive, they just provide foreigners/new immigrants with enough information to refer to public institutions in case of necessity and/or with regard to the medical topic presented in the texts themselves. Instances of reduction and simplification are also apparent in the format – a single multilanguage pdf brochure as opposed to different brochures in different languages –, in the linguistic strategies deployed (grammar and lexical simplification) and in the widespread use of visuals vs. writing.

The fact that the above trends have been seen to be more deeply at work in the 2010 corpus (see §4) might presumably mean that a first stage of the paper brochure transferred to the web is represented by the 2007 documents, while the 2010 corpus includes examples of the initial, natural evolution of genres once they have migrated to the digital medium, prompted by the gluey effects of Web 2.0 participation. This confirms that the e-brochure genre has been – and is likely still – developing specialised features depending on the sector (e.g. Italy's healthcare) and target users (e.g. the new immigrants). Additional monitoring of the *Servizio sanitario nazionale* (Italian National Health Service, INHS) websites over time is needed to record and interpret any further changes in online multilingual material.

References

Andreassen, Hege K. / Bujnowska-Fedak, Maria M. / Chronaki, Catherine E. *et al.* 2007. European citizens' use of E-health services: A study of seven countries. *BMC Public Health.* 7/1, 53.

Askehave, Inger 1998. *A Functional Interpretation of Company Brochures – from context to text.* Ph.D. thesis. The Aarhus School of Business.

Askehave, Inger / Swales, John 2001. Genre identification and communicative purpose: a problem and a possible solution. *Applied Linguistics* 22/2, 195-212.

Baru, Rama V. 2003. Privatisation, of Health Services. A South Asian perspective. *Economic and Political Weekly* 38/42, 4433–4437.

Berruto, Gaetano 1987. *Sociolinguistica dell'italiano contemporaneo.* Roma: La Nuova Italia Scientifica.

Bhatia, Vijay K. 1993. *Analysing Genre: Language Use in Professional Settings.* London: Longman.

Bonaguidi, Alberto / Valentini, Alessandro 2007. A proposito di impatto dell'immigrazione. Il caso dell'Emilia-Romagna. *Neodemos.it, Popolazione, società e politiche.* <http://www.neodemos.it/index.php?file=onenews&form_id_notizia=112>.

Catenaccio, Paola 2009. Promoting medical tourism in India: building identity and reputation in health tourism websites. In Garzone, Giuliana / Archibald, James (eds) 2009 *Discursive Identities in the Marketplace and the Media.* Bern: Peter Lang, 59-86.

Eysenbach, Gunther 2001. What is e-health? *J Med Internet Res.* 3/2, e20.

Fairclough, Norman 1995. *Critical discourse analysis: The critical study of language.* London, New York: Longman.

Fairclough, Norman 2003. *Analysing Discourse.* London: Routledge.

Fox, Susannah / Jones, Sydney 2009. The social life of health information. Americans' pursuit of health takes place within a widening network of both online and offline sources. *Pew Internet & American Life Project.* <http://www.pewInternet.org/Re ports/2009/8-The-Social-Life-of-Health-Information.aspx?r=1>.

Garzone, Giuliana 2007. Genres, multimodality and the world wide web: theoretical issues. In Garzone, Giuliana / Poncini, Gina /

Catenaccio, Paola (eds) *Multimodality in corporate communication: web genres and discursive identity*. Milano: FrancoAngeli, 15-30.

Grego, Kim 2010. The headache from hell: www.aspirin.com between marketing and science communication. In *Specialised Usages of English in Economic and Business Contexts*, numero monografico di *RILA – Rassegna Italiana di Linguistica Applicata*. Roma: Bulzoni 42/3, 141-157.

Grego, Kim / Vicentini, Alessandra 2009. The Revolution of Devolution: Issues of identity in Italian Public Healthcare Websites. *Paper presented at the 7th Interdisciplinary COMET (Communication, Medicine & Ethics) Conference*, Cardiff, Wales, 25-27 June 2009. Submitted for publication to *Communication and Medicine*.

Grego, Kim / Vicentini, Alessandra 2011. Holiday dialysis in Italy on the web: Multidimensional hybridization in institutional health communication. In Sarangi, Srikant / Polese, Vanda / Caliendo, Giuditta (eds) *Genres on the Move: Hybridization and Discourse Change in Specialized Communication*. Napoli: Edizioni Scientifiche Italiane, 393-406.

Grego, Kim / Vicentini, Alessandra forthcoming. The lexicon of ethics in medical discourse: selected national healthcare systems' webpages. In *Atti del convegno Lexicology and Lexicography of Domain-Specific Languages*, Università degli Studi di Palermo, 21-23 giugno 2007. Monza: Polimetrica International Scientific Publisher.

Heinze, Trisha 2004. *Informational Medical Brochures*. Final MA dissertation paper. Michigan State University. <https://www.msu.edu/~heinzetr/final2.html>.

Held, David / McGrew, Anthony 2003. *Global transformations reader. An introduction to the globalisation debate*. Cambridge: Polity.

Hesse, Bradford W. / Nelson, David E. / Kreps, Gary L. *et al.* 2005. Trust and sources of health information. *Arch Intern Med* 165/22, 2618-24.

Internet World Stats 2010. <http://www.Internetworldstats.com>.

ISTAT 2008, 2009. *Cittadini stranieri residenti per sesso e paese di cittadinanza al 1° gennaio 2008, 2009.* <http://demo.istat.it/altridati/rilbilstra>.

Journal of Medical Internet Research. <http://www.jmir.org>.

Kress, Gunther / van Leeuwen, Theo 2006. *Reading Images. The Grammar of Visual Design.* London: Routledge.

Maurais, Jacques / Morris, Michael A. (eds) 2003. *Languages in a globalising world.* Cambridge: CUP.

May, Steven K. / Cheney, George / Roper, Juliet (eds) 2007. *The debate over corporate social responsibility.* Oxford: Oxford University Press.

Meyerhoff, Miriam 2006. *Introducing Sociolinguistics.* New York: Routledge.

Ministero della salute 2010. <http://www.ministerosalute.it>.

Mooney, Annabelle / Sarangi, Srikant 2003. Click here for health information/advice: interaction pathways via the NHS Direct Website. In Crivelli, Benedetta / Rubinelli, Sara (eds) *Televisione, stampa e Internet tra medico e paziente.* Numero speciale di *Tribuna Medica Ticinese*, 21-27.

NHS UK 2010. <http://www.nhs.uk>.

Orizio, Grazia / Rubinelli, Sara / Schulz, Peter J. / Domenighini, Serena / Bressanelli, Maura / Caimi, Luigi / Gelatti Umberto 2010a. Save 30% if you buy today. Online pharmacies and the enhancement of peripheral thinking in consumers. *Pharmacoepidemiology and Drug Safety* 19/9, 970-6.

Orizio, Grazia / Rubinelli, Sara / Nava, Elisabetta / Domenighini, Serena / Caimi, Luigi / Gelatti Umberto 2010b. The Internet and health promotion: a content analysis of the official websites of Italian Public Health Authorities. *Journal of Medicine and the Person* 8/3, 121–127.

Permessi di soggiorno, ISTAT 2009. <http://demo.istat.it/altridati/permessi/index.html>.

Risager, Karen 2006. *Language and culture: Global flows and local complexity.* Clevendon and Buffalo: Multilingual Matters.

Shepherd, Michael / Watters, Carolyn 1999. The Functionality Attribute of Cybergenres. In *Proceedings of the 32nd Hawaii Interna-*

tional Conference on System Science (HICSS-32), 5-8 January 1999. Hawaii: Computer Society Press.

Slevin, James 2002. The Internet and Society: central themes and issues. In Brügger, Niels / Bødker, Henrik (eds) *The Internet and Society?* Århus: The Centre for Internet Research, 7-11.

Stranieri in Italia 2010. <http://www.stranieriinitalia.it/news/statistiche.html>.

Swales, John M. 1990. *Genre Analysis: English in academic and research settings*. Cambridge: Cambridge University Press.

Vicentini, Alessandra 2010. Language and marketing strategies in healthcare and corporate websites: The case of the American HHS. In *Specialised Usages of English in Economic and Business Contexts*, numero monografico di *RILA – Rassegna Italiana di Linguistica Applicata*. Roma: Bulzoni 42/3, 159-183.

Vicentini, Alessandra / Grego, Kim / Berti, Barbara / Bellini, Paolo / Orizio, Grazia 2012. Intercultural and Ideological Issues in Lexicography: A Prototype of a Bioethics Dictionary. In Facchinetti, Roberta (ed.) *English Dictionaries as Cultural Mines*. Newcastle: Cambridge Scholars Publishing, 247-264.

BETTINA MOTTURA

The Chinese Government Exploring Genres for Web-mediated Communication

1. Introduction

Identifying an internationally shared definition of the term 'e-government' goes far beyond the scope of this chapter, but it is relevant to point out that the Chinese academic literature on the subject attributes a quite extensive meaning to the term. Chinese e-government is defined as the set of all the electronically powered hardware or software tools used by public administration actors at all levels to perform their duties. The concept therefore encompasses companies' Intranets, institutional local area networks and websites published on the Internet.

This definition probably stems from the historical evolution of the project of e-government as it emerges in official documents, being therefore a political rather than a technical product. Wu and Li (2011: 21-30) describe the history of e-government in China as the creation of organs, norms, and technical instruments which first guaranteed office automation, then a flow of information within public administration offices and finally led to establish a regular flow of services and information between the government and society.

Since 2006 the Chinese Government has published its official online portal at the URL <www.gov.cn>. The <gov.cn> domain is common to all the websites of any public administration organ that goes online in the country (Wu/Li 2011: 27-28). This constellation of official pages is considered to be just one result of a broad set of policies enacted by the Chinese leadership since the beginning of the 1980s, aimed at building an e-government system.

This chapter will focus on the latest period of this historical process by presenting the results of work-in-progress research on the Chinese version of the Government portal as a source of updated textual products by the Chinese public administration. The first step has been the identification of the section of the portal where the public can find most of the texts used by the public administration to communicate online with society. The section has been selected on the basis of two preliminary considerations. The first is the assumption that in the second half of the previous decade in China a new emphasis was given to two main public administration functions: spreading information to the public and collecting feedback from society, i.e. a two-way communication flow. The second consideration regards the evidence that the National Government website has been identified as one of the crucial media to perform these functions. Section 2 illustrates the importance accorded to these aspects in official documents.

Based on these considerations, the analysis of the selected section of the Government website discusses the choice of linguistic resources recently adopted by the public administration in order to tackle the tasks described above. The underlying hypothesis of this chapter is that their selection by the public administration and their deployment have been guided by criteria that take specific account of the peculiarities of the web-mediated environment and its multimodal nature. Finally, from the samples of texts observed, I shall focus on three online interviews with the Prime Minister and discuss whether these texts provide evidence of an effective exploitation of the online medium by state organs.

2. The government informs and listens to people

Neither the function of informing people nor that of gathering feedback from them are totally new to the Chinese government. As far as the information function is concerned, China has had a long tradition of centralised control over the news, historically called 'propaganda system', a mechanism aimed at regularly selecting the materials to be published by the media, which involves both the Chinese Communist Party and state organs (Shambaugh 2007). Still, in the past the public administration was involved in producing the news only through a few powerful and highly specialised organs (e.g. the Xinhua News Agency). In parallel, the responsibility of the state apparatus for keeping its finger on people's pulse, or for doing 'organization work',[1] has always been shared with grass roots party cells and committees of the Chinese Communist Party, and with mass organizations (trade unions, women unions, etc.).

Nonetheless, the emphasis put on the importance of these two functions can be considered as to how it has affected recent political orientations. Nowadays in China the public administration at all levels is directly involved in creating and spreading a timely and updated flow of information about its role and the results achieved in performing it. The expression "service-oriented government 服务型政府" appeared for the first time in a top level leader speech in 2004.[2] Since then the concept has been repeated several times in official party or state documents, for example in the secretary general report to the 17th Chinese Communist Party Congress in 2007 and lately in the Premier's Report on the work of government to the National People's Congress in 2011 (Wang 2010: 6; Wen 2011b).

To guarantee a service-oriented government, communication is a crucial factor, both in terms of the state informing the citizen (of relevant news, newly adopted rules and regulations, etc.) and in terms of the government being updated on the people's needs and degree of satisfaction about the services received. The concept of service-oriented government clearly emphasises the central position of the citizen in government-society relations. In this respect, the expression

---

1 One of the Party's tasks is to organise and guide society, in order to achieve political goals through mobilisation.

2 Premier Wen Jiabao referred to this concept in a speech at the Central Party School in February 2004 (cf. Wu/Li 2011: 42).

seems tightly linked to a tenet of contemporary Chinese political discourse: the goal of "putting the people first 以人为本" (Li 2010).

In addition, in May 2008 specific *Regulations of the People's Republic of China on Open Government Information* became effective. The "disclosure of government information 政府信息公开" is depicted as a systematic project for administrative organs, bound to offer the citizens user-friendly access to all relevant government information (art.1). This policy, state the regulations, is meant to contribute to attaining a wide range of objectives (i.e. administrative efficiency, government transparency and effectiveness, socialist democracy).[3] It is aimed at creating a multiple-layered system of exchanging information between the public administration and the citizen. The central and local governments at all levels should appoint a number of spokespersons and regularly hold press conferences, give interviews, issue press releases, etc. Also, people can now ask the administration for information on any topic related to their personal interests.

In parallel, in 2009 the *National Human Rights Action Plan of China (2009-2010)* listed among the civil and political rights of the Chinese citizen the "right to be informed 知情权": "The Chinese government will make more efforts to keep the public informed of government affairs, and improve relevant laws and regulations, so as to guarantee citizens' right of information" (cap.II, art.5).[4]

Considering the above mentioned goals, an official government website seems to be a potentially very effective channel to support both ways of communication (government to society and citizen to government). Therefore, it is not surprising that both the *Regulations* and the *Action Plan* explicitly indicate the government websites at all administrative levels as crucial tools for their implementation.

3 The Chinese original can be found in State Council (2008); an English translation by the Yale Law School is available at <http://www.law.yale.edu/documents/pdf/Intellectual_Life/CL-OGI-Regs-English.pdf> (accessed September 2011).

4 The Chinese original can be found in Information Office of the State Council (2010); an English translation by the China Daily is available at <http://www.chinadaily.com.cn/china/2009-04/13/content_7672483_2.htm> (accessed September 2011).

3. The government's website as a showcase of public administration genres

The government's website displays a wide selection of texts belonging to different genres. As a whole it shows an entire genre system: "[...] a complete set of discursive forms that are invoked by all the participants involved in a professional activity" (Bhatia 2004: 55). On the homepage, we can find links to at least four different kinds of sections, which can be identified through the genres they exploit to perform their functions. Some of the genres in use online are conventionally linked to the administrative work of state organs. For instance, the publication of rules and regulations, government reports, white papers and official documents provides some good examples of how the website is used as a support for the circulation of traditional bureaucratic genres. Also, in terms of their electronic dissemination, in terms of layout, structure, form and functions these texts do not deviate from the norm of the genres they belong to,[5] so they can be considered a mere 'electronic equivalent' of their paper forms (Crystal 2006: 204).

Other genres belong to the realm of self-promotion. Published with either a descriptive or an informative intent, these texts offer the citizen a positive or reassuring representation of government institutions, as efficient bodies seeking to improve the country's conditions and solve problems. Examples of these genres can be found in the sections presenting leaders' profiles and reports on their daily activities, country facts and figures, government structure and functions, and state of the art in administrative reforms.

A third category of genres hosted by the government website can be situated in a news report framework. Under the press releases label, we can find texts recording either government spokespersons'

5 Some of them, for example, are realisations of the genres codified by the *Method for the Discipline of the Official Documents Issued by the Country Administrative Organs*, a document adopted in 2000 by the State Council of the People's Republic of China (Ruihuan 2009: 386-396) in order to standardise the genres in use in the public administration.

statements or press conferences with questions and answers. The section displays an archive ordered by date, and recent posts are classified according to the issuing body. The use of these information mechanisms and genres is an outcome of the implementation of the above mentioned open government information policy. Last but not least, on the government portal there are pages devoted to offering services to citizens.

4. A section of the government's website focussed on online communication with society

In consideration of the complexity of the Chinese government's website and the overlapping functions performed by a number of its sections, the analysis will now focus on a part of the website entitled 'Interacting with government affairs 政务互动'. This choice is based upon the explicit goal stated in this title and on the relevance of its content to this discussion, given that the first target of the analysis of the website is to identify some of the genres adopted by the public administration to communicate online with society.

In light of the definition of genre put forth by Swales (1990: 45-58) and Bhatia (1993), the genres under examination here share the primary purpose of enhancing a two-way communication between the Chinese government and the public at large. As pointed out by the policies illustrated above, in addition to the goal of informing, they are also likely to serve the purposes of administrative efficiency, government transparency and effectiveness. These texts also aim to contribute to moulding public opinion (Zhang 2010: 50) and increasing the legitimacy of the government to rule the country (Shambaugh 2008: 106-111). The 'Interacting with government affairs' section shows five different sub-pages: policies interpretation 政策解读, online interviews 在线访谈, suggestions and remarks to the government 建言议政, collecting opinions 意见征集 and online surveys 网上调查.

4.1. The government's voice

In 'Policies interpretation 政策解读', entries lead to government explanations of policy goals or targets. These texts display some of the traditional features of the official documents. The body publishing the explanation is identified by means of its official name, and the date and the original location of the online post are clearly marked. The title of each document defines its main topic and the text is divided into paragraphs (sometimes structured as Q&A). The publishing organ is identified as the author of the text, but in most cases it attributes the responsibility of its interpretation to a specific public officer. The expert is always defined by his/her professional role, but is not always referred to by his/her full name (generally the higher the position, the more likely it is to find the person's name).

The 'Suggestions and remarks to the government 建言议政' section contains reviews, comments, and commentaries mostly published by the Xinhua Press Agency. These texts are press releases by the national press agency, imported into the government's website. They therefore display the date of issue, the author's name (a journalist usually), a title identifying the main topic in the document and expressing some sort of evaluation on the matter. As the text is produced by a journalist and usually contains a comment on the topic addressed, the government offices do not appear as the source of information, but as the object of analysis. Nonetheless it is crucial to remember that actually the Xinhua Press Agency is one of the organs of the State Council, therefore it represents a journalistic point of view which stems from the government: in this context, the professional identity of the journalists is not matched by critical independence.

The 'Online interviews 在线访谈' section contains a list: each entry is a title which identifies the public administration's organ participating in the interview and the main topic discussed. The most recent post is dated end of February 2012. All titles are links referring to further pages, where one can find pictures, excerpts from the interview, its full text and further links to background materials as well as the original text of the questions or comments by the public.

On these pages the first picture at the top shows two or more people sitting at a table talking to each other with a laptop computer opened in front of them. They play different roles: the interviewer is a journalist and the interviewee is a state organ's spokesperson or member of staff. The picture is followed by a caption identifying the time, the host and the topic of the conversation; and inviting the 'People of the web 网民' to ask questions and participate in the discussion.[6] The background is China's national flag, and a faint image of the south gate of the Forbidden City, both of which are symbols of the country's unity and state's power.

The whole section is therefore devoted to a government representative answering people's questions on relevant topics selected by the anchorperson. The answers are published as transcripts on the website. The communication flow is therefore mediated both by the computer (in receiving questions and answering) and by a journalist (in selecting topics and typing answers). This double mediation intensifies some aspects of web interaction, such as the use of the written medium and the time gap in interaction.

4.2. In search for people's voices?

The 'Collecting opinions 意见征集' and 'Online surveys 网上调查' sections fulfil a common function: to publish legal or policy texts in their draft form and call for the public's response to their content before formally adopting them. Both channels share the same pattern, with slight differences. The draft of the legal/policy text is displayed (directly on the page in .htm format or through a link that refers to an attachment in .rar or .pdf formats) in its formal shape, associated with

6 The expression 'People of the web 网民' is broadly used in China as an equivalent of 'netizens'. If we compare the two words an interesting difference emerges: one side refers to Internet users as citizens, the other as people. Both concepts appear to be deeply rooted in political culture and history.

the typical features of administrative documents.[7] The draft is embedded in a textual frame where the issuing organ explicitly addresses the general public or specific stakeholders, and asks for comments on the measure.[8] This part of the page describes the feedback-giving mechanism; it usually indicates a telephone number or email address (associated to the name of the contact person) or an online procedure to be performed on a parallel website.[9] The linguistic features of the frame text are homogeneous with those of the draft: the language is formal and standardised, the authorship is clearly designated, and the date of publication and a deadline for feedback are indicated.

Nonetheless, there is at least one discordant feature which sheds light on a possibly less organised and normalised working context behind the scenes: when an email address is provided for feedback, it is not always linked to a mailbox hosted by the official government platform. This idiosyncrasy demonstrates that either not all the government organs provide their employees with an official email address, or that civil servants prefer to use other internet service providers for the task, or maybe the individuals in charge of handling the public's feedback are not always government employees. In neither of the two sections does there seem to be any access by other netizens to people's feedback. As far as I can see, today the bottom-up communication flow is still limited to single actors talking to the government (Jiang/Xu 2009), while no open discussion arenas seem to be available.[10]

---

7 It is identified by a title, the issuing organ's name, a date, and most of the time a reference number. The text is divided into paragraphs and often into numbered sections.

8 In this respect there seems to be a difference between the two sections: in 'Collecting opinions' most texts call for other administrative organs' feedback, while in 'Online surveys' the intended audience is society or people.

9 The online access to feedback is mostly offered by the Legislative Affairs Office (to access the process, netizens need to register first); other organs usually prefer indicating an email address or a telephone number.

10 The most recent posts in both sections date from February 2012.

5. Prime Minister's online interviews: exploring new communication boundaries?

In creating the new communication channels between the state and society described in the previous paragraph, the relevance of the web as a medium of communication is to be questioned. On the one hand, it is self-evident that there is an effort to benefit from some of the affordances of the online medium. As mentioned before, the Chinese government is exploiting the Internet to achieve at least three targets. The website overall facilitates a timely and broad distribution of information. The various pictures and voices represented on the website help to highlight the institution's internal plurality of faces and its vitality. The call for public feedback testifies to state organs' willingness to include other subjects' voices in the decision-making process, reinforcing a process of co-operation and strengthening the legitimacy to govern.

On the other hand, as already noted, at least two other qualities of web-mediated communication seem not to have been taken into consideration. No fully synchronous activities seem to have ever been realised on the 'Interacting with government affairs 政务互动' page. At the same time, there is no sign of a shared arena to foster a peer-to-peer exchange of ideas or opinions; rather, the communication flows always seem to be vertically oriented (government-to-citizen or citizen-to-government).

In order to confirm this view, I will analyse one of the communicative events described above: the online interview, which can be considered a genre strictly dependent on the online medium, insofar as it cannot be realised without an Internet access. Because of the variety of personalities involved in online interviews and the number of matters discussed, for the present work I focus only on the analysis of interviews delivered by Prime Minister Wen Jiabao published in the 'online interviews' section. These three texts date from February 2009, 2010, and 2011; and consist of about 17,500 Chinese characters in the 2009 interview, 18,000 in the 2010 interview and 19,000 in the

2011 text.[11] The choice is based on the role of the Prime Minister in leading and representing the government, which possibly guarantees a high quality involvement of the institution and the media in the communication process (both in technical and strategic terms).

5.1. The web-mediated communication process

The three web pages hosting the selected interviews by Mr. Wen are graphically homogeneous and similar to other online pages featuring interviews. The overall graphic structure of the page is simple and clear: it is divided into a few boxes with text or pictures. The prevailing colours are red (the heading banner, the section title frames and the pictures' background), white (the page background) and black (the script).

The general effect is one of straightforwardness: the black writing on a white background is very clear and easy to read. As far as colours are concerned, the main issue seems to be the use of red. In China the colour is traditionally linked to good fortune and is politically associated with the Chinese Communist Party.

On the top there is a space for interchanging pictures and one for a video of the interview.[12] Invariably, these pictures show an informally dressed Prime Minister, seated behind a big table with two portable computers and two tea cups. This sort of mug with a lid is often seen on tables at high-level talks and therefore can be considered a symbol of formality. But at the same time they convey a cosy feeling, as offering a cup of tea is a very common act when receiving guests in most Chinese houses.

---

11 From now on any references to linguistic features of the texts refer to the full texts of the interviews in Chinese published respectively at the following addresses: <http://www.gov.cn/zlft/>; <http://www.gov.cn/zlft 2010/>; <http://www.gov.cn/zlft2011/> (accessed February 2012). No new interview with Wen Jiabao has been posted in the first two months of 2012.

12 In February 2012 I did not manage to play any of the videos.

Each year, on Wen Jiabao's left side sits a journalist from the Xinhua News Agency, in all three cases a woman. The young, pretty and respectful journalists at his side emphasise year after year Mr. Wen's wisdom (he is older than the journalist interviewing him) and prestige (the woman overtly admires him). The smaller pictures at the bottom of the page show the leader interacting with different people, often smiling. In these pictures the main message seems to be: the Prime Minister is very active in meeting citizens, and he is receptive and friendly.

Different page sections contain links referring to: a list of highlights from the interview, the complete transcript of the interview, and the record of people's questions and comments. Although the number of questions posted by the public decreased from 303,768 in 2009, to 239,778 in 2010 and 105,130 in 2011, the length of the interview transcript has expanded from year to year.

Working through these texts, it is possible to retrace the process through which the interviews are constructed. The public sends questions through a virtual forum or by sms, the journalist (or, more likely, a team of journalists)[13] conducting the interview selects the most frequent or relevant topics and presents them to the Premier. Then Wen Jiabao discusses the subjects one by one.

The interviews are structured in the customary shape of questions and answers between each anchorwoman and the Premier. Graphically, therefore, in the three texts the script of the interviews represents the form of a conversation between two or more voices. Each speaking subject's voice is introduced by her/his title or name and the colons, followed by her/his contribution to the conversation. The Premier, the anchorwoman and individuals from the public take turns in expressing their views.

In the alternation pattern (Xing 2006: 220) of these conversations there is a patent imbalance in turn frequency and size, and a kind of regularity in turn order. Proportionally Wen Jiabao's contributions, which usually follow a brief introduction by the journalist, prove to be

13 This hypothesis stems from comments expressed by anchorwomen, who explicitly declare that they are summing up different people's requests when asking the Premier a question.

more frequent than those of the others and longer. The anchor-woman's voice is much more present at the beginning of the conversation, then it tends to appear rarely but regularly with short turns. The public, represented by a number of individuals,[14] directly contributes only a few times, most of the turns being introduced by the journalist.

These texts can be considered complex discourse media (Crystal/Davy 1969: 70): Mr. Wen's oral performance stems from written questions from the public and is rendered into written form to be published online. It is self-evident that, as highlighted by Crystal (2006: 31-52), this written form of online interaction "relies on characteristics belonging to both sides of the speech/writing divide" (Crystal 2006: 31-52).

Let us compare the interviews with the main features of oral conversation. In the transcripts there are no traces of the inexplicitness or of the disfluencies of conversation (Crystal/Davy 1969: 102-104). There is a lack of simultaneous feedback and the communication rhythm is slower than in face-to-face interactions (Crystal 2006: 32-33). Also, the role of the anchorwomen[15] in selecting and organising the topics randomly addressed by the public overcomes the usual lack of planning characterising spoken discourse (Crystal/Davy 1969: 103). Moreover, relying on the description of basic Chinese conversation modes outlined by Xing (2006: 216-227), it is possible to identify some peculiar linguistic features of conversations in the texts. The transcript shows some examples of the common initiation formulas (Xing 2006: 218).[16] It is immediately clear that the overall register is quite formal, especially between the journalists and the Premier, and still both authorities address the public by calling it "friends". Interjec-

14 The public is represented by individuals, and still each of the speaking voices identifies a social group or stratum: peasants, migrant workers, university students, factory workers, etc.

15 I believe this selection and organization of discourse topics is likely to be the product of the behind-the-scene work of an entire team, but there is no explicit proof of that (besides the obvious participation of technical support in the interview).

16 The anchorwoman says: "Dear friends, hallo everybody. 各位朋友，大家好", "Premier Wen, how do you do. 温总理，您好!"; the Prime Minister answers: "How do you do? Hallo to the Internet people! 您好。网友们好。".

tions (Xing 2006: 224-227) are not frequent in these texts. For example, in the 2011 interview there are: one occurrence of *ba* 吧 "used to solicit agreement"; three of *a* 啊 "used to express exclamation"; and three of *ne* 呢 "used to express coherence". All three texts end with Premier Wen Jiabao and the anchorwomen waving good-bye to the public, as if they had really met.

5.2. Relevant content pattern and linguistic features of Wen Jiabao's voice in the online interviews

Wen Jiabao online interviews set a communication flow between the government and society, and are therefore characterised by some distinctive features of this kind of exchange (Pellat/Liu 2010: 109). First, the relationship between the speaker and the public is unequal, there is a patent imbalance of power between the two parts. Second, the leaders tend to illustrate any matter under the best possible light or at least they depict the future as positive, and all disasters or defeats are explained or justified. Third, there is a constant appeal to patriotic sensitiveness by promoting a specific view of the nation. Fourth, the language is inclusive.

Since the Premier's contributions to the online conversation analysed here are the core of the communication flow in terms of length and relevance, I shall primarily focus on Wen's part of the speech and highlight some of its recurring patterns. The first such pattern is presenting credentials and establishing reputation. At the very beginning of each interview the leader refers to his own position and responsibilities. He talks about himself as the Premier and states for example: "being the Chinese Premier is difficult, one has to bear huge responsibilities 做中国的总理难，责任重大 (Wen 2009: 15:09)"; "I have been in Zhongnanhai[17] for 25 years now, I can be considered a no-vacation-time man, 我在中南海25年，可以说是一个没有节假日的人" (Wen 2010: 15:02); "My term of office will expire in two-years' time,

17 Zhongnanhai is the compound of Chinese national leaders' residences in Beijing; living there means holding a high-level post in the Chinese communist party or in the state administration.

I assume this two-years' work won't be any easier than that of each of the last eight years, on the contrary it will be much more difficult 我的任期还有两年，我以为这两年的工作不比前8年任何一 年要容易，反而要艰巨得多" (Wen 2011a: 09:02). Besides, he makes a constant use of references to national history or quotations of the country's forefathers, which place him firmly in the long tradition of national leaders. Being the heir of noble ancestors and emphasising continuity are traditional ways of affirming political legitimacy in China.

Addressing a number of social and economic issues raised by the questions from the public,[18] Wen Jiabao proves to be aware of the existing problems and ready to assess the real situation. First he regularly quotes numbers and percentages to demonstrate his expertise. The formal know-how is then backed by frequent accounts of the Premier's direct contact with people when he travelled through the country: "a few years ago I went to Changchun for an inspection visit to a hospital 前几年有一次我在长春出差， 到一所医院去视察" (Wen 2009: 15:48); "I went to the drought region of Guangxi to spend New Year's Eve with the people and to check the damage caused by the drought 我到广西旱区，我在那里一边和群众过年，一边了解旱情" (Wen 2010: 15:06); "when I went to Zhejiang to make a survey last year [...] 去年我在浙江考察的时候" (Wen 2011a: 09:22).

The stories of common people or the personal anecdotes Mr. Wen tells appeal to the public's emotions and are mostly linked to positive social values such as family cohesion, parenthood, fairness, trustworthiness, benevolence and respect. These pictures display shared identities, hopes, difficulties and griefs. The Premier himself expresses personal feelings.

In parallel, the widespread use of the personal pronoun *we* 我们 and the possessive adjective *our* 我们(的) unveil the use of an inclusive language. In the texts, there are at least two different extralinguistic referents for each of the two words: sometimes it is the Chinese people/nation, in other cases it is the Chinese leadership/government.

18 In 2011 some of the issues were: happiness, commodities prices, residence permit and children's education, higher education, energy savings, real estate, corruption, etc.

The first option prevails in terms of occurrences. In both cases Wen Jiabao considers himself part of the group. When referring to the Chinese people he places himself at people's side, when pointing at the governance of the country he usually takes the responsibility for specific commitments. The reader always gets the impression of being important, either as involved in a collective action or as the witness of a pledge for a better future.

6. Conclusions

In Section 3 of this chapter I indicated three possible functions of the textual genres represented on the <www.gov.cn> website: the execution of administrative functions, self-promotion, and informing the general public or guaranteeing public services. If we consider the online interviews with Wen Jiabao analysed in the previous section as realisations of a genre, they are undoubtedly to be placed under the self-promotion label.

The interview is well established as a journalistic genre and can be described as a dialogical situation where two persons, being in the same place, take turns in speaking, and where, due to their different social status, one is entitled to ask questions and lead the conversation (Charaudeau 2005: 179). In the realisations of this genre, the interviewer's role is that of selecting the topics to be discussed and convincing the interviewee to be honest and thorough in answering. Charaudeau identifies the political interview as a sub-genre where the host is an important politician. In this case, the main objective of the journalist should be to smoothly force the authority to deliver as much information as possible and reveal her/his hidden intentions in the field of discussion addressed (Charaudeau 2005: 179).

These definitions only partially apply to the Chinese case study analysed. As I highlighted in Section 5, the basic pattern of the interview is confirmed, but there still are some differences in participants' roles and in the final textual realisation. First, in the interviews with the Premier the scope of the journalist's leading role seems to be quite limited. The three anchorwomen do select the topics to be discussed, but only introduce the subject and never interact with Wen Jiabao's answers. This prevents any possibility of clarification or of broadening the authority's contribution. Therefore, it is the Prime Minister who actually defines the scope of the conversation. Second, the audience in this case is not physically present (as assumed by Charaudeau), but only virtually present. This peculiarity is due to the online environment.

These two specificities of the texts analysed contribute to confirming the self-promotion hypothesis. The limited interaction between the journalist and the leader gives the Premier time and space to present situations in the best possible perspective so as to enhance the government's prestige. While the presence of the audience, mediated through the web, contributes to representing the minister as being willing to confront the general public and listen to its point of view, at the same time it guarantees the possibility to select and (again) to control the communication flow. Ultimately, it is Wen Jiabao who is guiding the process, therefore he can easily emphasise the positive aspects of any situation he talks about and reinforce the legitimacy of his government to be in charge. And in the meantime he fosters the country's social cohesion and the sense of a shared identity.

References

Bhatia, Vijay K. 1993. *Analysing Genre. Language Use in Professional Settings*. London: Longman.

Bhatia, Vijay K. 2004. *Worlds of Written Discourse. A genre-based view*. London: Continuum.

Charaudeau, Patrick 2005. *Les médias et l'information. L'impossible transparence du discours*. Bruxelles: De Boeck-Ina.

Crystal, David / Davy, Derek 1969. *Investigating English Style*. London: Longman.

Crystal, David 2006. *Language and the Internet*. Cambridge: Cambridge University Press.

Information Office of the State Council 2009. 国家人权行动计划 (2009-10) (National Human Rights Action Plan of China (2009-10) <http://www.gov.cn/jrzg/2009-04/13/content_1283983.htm>.

Jiang, Min / Xu, Heng 2009. Exploring Online Structures on Chinese Government Portals: Citizen Political Participation and Government Legitimation. *Social Science Computer Review* 27, 174-195.

Li, Qing 2010. *科学发展观视域中的'以人为本' ('Put the People First' in the Boundaries of the Scientific Development Perspective)*. Beijing: Shishi Chubanshe.

Pellat, Valerie / Liu, Eric T. 2010. *Thinking Chinese Translation*. London: Routledge.

Ruihuan, Ji 2009. *常用公文写作方法与技巧 (Methods and Techniques to Write Common Official Documents)*, Beijing: Zhongguo Renshi Chubanshe

Shambaugh, David 2007. China's Propaganda System: Institutions, Processes and Efficacy. *The China Journal* 57, 25-58.

Shambaugh, David 2008. *China's Communist Party. Atrophy and adaptation*. Berkeley: University of California Press.

State Council 2000. 国家行政机关公文处理办法 (Method for the Discipline of the Official Documents Issued by the Country Administrative Organs), in Ji Ruihuan 2009. *常用公文写作方法与技巧 (Methods and Techniques to Write Common Official Documents)*, Beijing: Zhongguo Renshi Chubanshe, 386-396.

State Council 2008. 中华人民共和国政府信息公开条例 (Regulations of the People's Republic of China on Open Government Information). In Policy and Legislative Office of the Ministry of Personnel (ed.) *人事工作文件选编* (*Selected Work Documents of the Ministry of Personnel*). Beijing: Zhongguo Renshi Chubanshe, 14-23.

Swales, John M. 1990. *Genre Analysis*. Cambridge, UK: Cambridge University Press.

Wang, Changsheng 2010. 融合与创新：电子政务发展新阶段 (Integration and Innovation: Advancing the E-Government to a New Stage). In Wang, Changsheng / Jie, Xiaoping (eds) *中国电子政务发展报告 (2010) (Annual Report on China's E-Government Development (2010))*. Beijing : Social Sciences Academic Press, 1-9.

Wen, Jiabao 2009. 温家宝总理与网友在线交流。2009年2月28日，北京 (Premier Wen Jiabao Communicates with Online Friends. Beijing February 28th, 2009). <http://www.gov.cn/zlft/>.

Wen, Jiabao 2010. 温家宝总理与网友在线交流。2010年2月27日，北京 (Premier Wen Jiabao Communicates with Online Friends. Beijing February 27th, 2010). <http://www.gov.cn/zlft2010/>.

Wen, Jiabao 2011a. 温家宝总理与网友在线交流。2011年2月27日，北京 (Premier Wen Jiabao Communicates with Online Friends. Beijing February 27th, 2011). <http://www.gov.cn/zlft2011/>.

Wen, Jiabao 2011b. *政府工作报告。* 2011年3月5日 (Report on the Work of Government. March 5th, 2011). <http://www.china.com.cn/policy/txt/2011-03/16/content_22150608.htm>.

Wu, Jiang / Li, Zhigeng 2011. *电子政务与服务型政府建设 (E-government and the Construction of a Service Oriented Government)*. Beijing: Guojia Xingzheng Xueyuan Chubanshe.

Xing, Janet Zhiqun 2006. *Teaching and Learning Chinese as a Foreign Language*. Hong Kong: Hong Kong University Press.

Zhang, Shuhua 2010. *网络民意与公共决策：权利和权力的对话 (Online Public Opinion and Public Policies: a Dialogue between Rights and Authority)*. Shanghai: Fudan Daxue Chubanshe.

CHIARA DEGANO

Argumentative Genres on the Web: The Case of Two NGOs' Campaigns*

1. Introduction

The migration of existing genres to the web and the creation of new web-native genres have been a major element of genre variation as a consequence of the effect produced by the medium on the text (Askehave/Ellerup Nielsen 2005). In particular the loss of linearity brought by hypertext has drawn the attention of scholars, being the most macroscopic change at the level of textuality introduced by website writing.

All text types are potentially affected by such a loss of linearity, but it is the argumentative text that risks being more affected, considering the importance attributed universally to *dispositio,* i.e. the ability to arrange arguments in an effective order, since the time of classical rhetoric. While traditionally responsibility for *dispositio* has been conceived as resting exclusively with the author, this is not the case in hypertexts where the user contributes to articulating the text (Garzone 1997; Lemke 2006), thus preventing the author from imposing a pre-arranged sequential order. In light of these considerations, this chapter investigates the influence of the web medium on computer-mediated argumentative texts, analysing in particular NGOs' campaigns on their websites. This is a genre which is overtly argumentative and has known an unprecedented boost with the spread on the Internet, which has given activist groups the chance of reaching a much broader audience.

* This chapter is part of the MIUR-funded research project PRIN 2007 "Tensioni e variazioni nei generi settoriali inglesi / Tension and change in English domain-specific genres" prot. 2007JCY9Y9.

The effects of the hypertext environment on textuality have been the object of quite extensive research not only within the field of linguistics (Fritz 1998; Tosca 2000; Bolter 2001; Kress/van Leeuwen 2001; Askehave/Ellerup Nielsen 2005; Garzone *et al.* 2007) but also of computer science. However, as pointed out by Carter (2000) with respect to the latter, most research has focussed on informational and literary hypertexts, leaving argumentative hypertexts largely unexplored. As for argumentation theory, the concern for computer-mediated communication has generally involved the mutual relevance of argumentation and artificial intelligence (cf. for example Walton *et al.* 2008: 38-40; Prakken 2005) and the conditions for argumentation created by protocols for discussion and debate on the web (cf. among others Jackson 1998; Weger/Aakhus 2003; Aakhus/Jackson 2005; Lewiński 2010).

Relying on the existing studies of argumentative hypertexts (particularly Carter 2000, 2003), as well as on linguistic research on web-mediated discourse (in particular Garzone 2007 and Askehave/Ellerup Nielsen 2005), this chapter analyses two cases of NGOs' campaigns: Baby Milk Action, against the marketing of infant formula with special regard for developing countries, and Greenpeace's campaign against genetically modified food. While their specific content is not crucial for the investigation carried out here, the fact that both campaigns target commercial interests in the nutrition industry makes the two cases all the more comparable, reducing the incidence of differences which might be dependent on contextual factors.

The analysis will address issues connected with the loss of linearity in hypertexts, with special regard for its impact on textual coherence, and the consequential loss of the writer's control on the order of arguments (*dispositio*). While insights from Argumentation Theory, and particularly the pragmadialectic approach (cf. among others van Eemeren *et al.* 1996 and van Eemeren 2010) will be used in order to account for the macrostructural organization of hypertexts, categories of text grammar (Werlich 1983) will be applied to micro-levels of linguistic codification within single nodes, thus integrating the holistic concerns of genre analysis with more fine-grained linguistic perspectives.

2. Theoretical background

With the spread of the Internet a number of genres that had traditionally been transmitted through print or speech 'migrated' to the Web, and in doing so underwent a process of transformation under the influence of the new medium[1]. At the same time a host of new electronic genres were developed, responding to new communicative contexts made available by the web itself (Crowston/Williams 1997).

In light of these changes, scholars from different disciplines stretching from communication and linguistics to computer science started to investigate, each from their specific perspective, the features of web-mediated genres, coming to the conclusion that the theoretical tools available for genre analysis (Swales 1990; Bhatia 1993) should be revised and updated in order to accommodate the peculiarity of the new digital genres (Askehave/Ellerup Nielsen 2005; Garzone 2007).

Askehave and Ellerup Nielsen proposed to integrate the medium dimension in the Swalesian traditional model of genre, identifying multimediality (the combination of different 'sub-media' into one format) and hypertextuality (the 'key medium' used on the Web to connect texts) as the properties of the Web that most influence web-mediated texts (Askehave/Ellerup Nielsen 2005: 4). Both these properties affect the way in which a text is received, favouring a non-linear reading process, but it is hypertextuality in particular that Askehave and Ellerup Nielsen bring to bear on Swales' three-level model. Drawing on Finnemann's distinction between the reading mode and the navigating mode (1999) – where the former corresponds to traditional sequential reading while the latter allows the user to choose his/her reading path – they suggest incorporating this dual dimension in Swales' model. The three original constituents – communicative purpose, moves and rhetorical features – should then be considered from the twofold perspective of the reading and the navigating mode. While

1 In fact, as Garzone (2007: 18) points out, the web should rather be considered an environment (namely a Hypermedia Computer-Mediated Environment, drawing on Hoffman/Novak 1996), than a medium, as it integrates various media, resulting in a defining property of the Web, i.e. multimodality.

in the reading mode the three levels remain unvaried, as the user basically reads the text as if it were printed, in the navigating mode the category of moves is replaced by that of links, as it is links that determine the direction of reading, providing access to other related texts, according to the choice made by the reader.

2.1. Arguments on the web

If the navigating mode is potentially disruptive of any superimposed linearity for whatever type of discourse, its effect can be all the more problematic for argumentative discourse. In informative discourse, for example, the loss of linearity inherent in hypertexts is compensated for by the affordance of "instant and easy access to volumes of information, easy navigation, easy search capabilities and easy retrieval of data" (Carter 2000: 85). In argumentative discourse, on the other hand, it is the logical sequence of reasoning that is fundamental, and selecting and arranging arguments in the best possible order (*dispositio* in classical rhetoric) has traditionally been the responsibility of the author.

Although modern argumentation has partly shifted away from the classical concern for the right order of arguments functional to winning a debate, placing greater emphasis on the rhetorical goal of choosing the right arguments for an intended audience[2] (Perelman/ Olbrechts-Tyteca 1969), order remains a tenet of argumentative discourse, if only for what concerns textual coherence. Suffice it to compare the different textual structure of news stories and editorials: the former, whose primary function is informative, can easily be edited by rearranging the order of paragraphs and even cutting them according to page composition constraints, a need catered for by the instalment structure (van Dijk 1988: 15). On the receiving end of the continuum, it is contemplated that readers can abandon the story without finishing it, or may just skim through information, but, provided they reach the end of the lead, the gist of the story will have been conveyed. Editorials, on the other hand, which are eminently argumentative, are ac-

2 This actually introduces another source of difficulties for argumentation on the web, since it is not possible to know in advance who the addressee is.

knowledged a different status: the textual structure imposed by the writer cannot be so easily disrupted in the editing phase as the texture created by cohesive devices (in the Hallidayan sense, which includes both De-Beaugrande/Dressler's standards of cohesion and coherence, 1981) is much more closely knit and intricate. By the same token, readers are expected to follow the author's reasoning in its unfolding, without departing from the linear reading modality.

On the Web, due to the possibility for the user to shift from the reading to the navigating mode, the writer has lost the exclusive control over the arrangement of arguments, thus raising questions as to the functioning of arguments in this environment. Closely associated with the writer's loss of control, allowing what Landow (1992) calls 'random access' to the text, is fragmentation. In Carter's (2003:7) words:

> Unlike textual units – paragraphs and chapters – that depend on neighbour text units to create coherence, these self-contained nodes are like islands of meaning, no longer relying on other nodes for context.

On this ground scholars acknowledge that the web medium obliges the user to revise the communicative strategies used for argumentation, starting from the necessity to "build, in place of a single argument, a structure of possibilities" (Bolter 1991: 119). Relying on previous studies (Hovland 1957; Kolb 1997; Bernstein 1998) Carter presents some of the techniques that could be used to cope with the loss of linearity in digital argumentation, which essentially involve two superordinate strategies: avoiding fragmentation in the main parts of the argument (i.e., following Toulmin's [1958/2003] model of claims, grounds, warrants and data); and trying to orient the user in the 'reading' process. The latter effect can be achieved by arranging the material in such a way as to suggest a certain order of access to the contents of the webpage (including the links to lower levels in the website structure), through the exploitation of primacy (the first thing the user encounters), recency (the last thing encountered) and repetition effects (Carter 2003: 8).

So far the discussion has focused on elements of discontinuity between traditional and digital argumentation. However, there are also some points of contact, which actually allow the researcher to apply existing traditional models to the analysis of argumentation on the web. A first contact point can be identified in the similarity between the real use of arguments in everyday life – as opposed to formal debates – and arguments on the web. If one observes how people argue in the real world (the scope of investigation of informal logic), their arguments will appear "digressive, [...] repetitive, ill-organized, incomplete, and multi-functional" (Johnson/Blair 1978). The second point of contact is to be found in the analytic process commonly referred to in the literature on argumentation as the reconstruction of argumentative discourse (cf. for example van Eemeren *et al.* 1993).

The pragmadialectical approach makes the point quite clearly by arguing that the four stages of the model for a critical discussion (opening, confrontation, argumentation and conclusion) are not necessarily realized in that order in real world occurrences of argumentative discourse. This is the reason why the original text has to be, so to say, 'deconstructed' with a view to identifying text units functioning as standpoints and arguments, irrespective of their position within the physical layout of the text. In this way a new linearity is constructed by the analyst on the basis of the coherence that underlies the argumentative reasoning, often disregarding the original linearity imposed by the author. Once again, recourse to editorials can illustrate this point: it is often the case that leading articles and comments open with a topic which is not immediately related to the core object of debate, which is disclosed later on in the text. If an analyst is to reconstruct the critical discussion carried out in the text, the first part of the article will have to be ignored in order to allow a streamlined structure to come to the surface.

The pragmadialectical theory proves a valuable frame of reference to this chapter's ends also for the attention it devotes to the rhetorical component of argumentation (Houtlosser 2006, van Eemeren 2010). Traditionally argumentation and rhetoric have been perceived as antithetical, with the former interested mainly in the logical soundness of reasoning put at the service of the pursuit of truth, and the latter associated with the less noble art of winning a debate, with scant or no consideration at all for truth. The pragmadialectical approach, on the other hand, has integrated the concern for logical soundness of argumentation with the pursuit of rhetorical effectiveness in the notion of *strategic manoeuvring*. This amounts to the efforts arguers make to pursue their ends along three dimensions: the selection of arguments, adaptation to the audience and presentational choices, and it is precisely in this latter dimension, which pertains to the linguistic 'stuff' of which argumentative moves are made, that the potential intersection between genre analysis and the pragmadialectical approach resides.[3] Recourse to the systematic pragmadialectical model of argumentation can put the analysis of argumentative genres on a sounder organizational footing, while taking into due consideration the rhetorical and discursive strategies deployed to realize genres' specific constitutive moves.

To this end, categories of Critical Discourse Analysis – CDA – (cf. among others Wodak/Meyer 2001) can also be brought to bear for the fine-grained description and interpretation of the rhetorical devices used to realize moves, with special regard for implicit meanings, like presupposition (cf. Bekalu 2006; van Dijk 2006) and intertextuality. While not dealing specifically with argumentation, nor with the hypertextual environment, CDA proves a useful complementation to the main approaches adopted for this study, not only for its attention for the micro-level of linguistic choices, used to construct a certain representation of reality, but also by virtue of its conception of language as social action. Such an assumption is particularly fine-tuned with the genre of NGOs' campaigns, characterized as they are by their social stake, addressing issues that are, or are presented as, socially relevant, and usually in opposition to some form of hegemonic power, be it political or economic.

3. Study design

Taking into account the structural constraints as well as the affordances characterising computer mediated communication, this chapter sets out

3 A further element of convergence can be found in the interest the pragmadialectical approach has recently developed for institutional discourse – a core concern of genre analysis – which is investigated with a view to determining how contextual factors influence argumentative practices in specific contexts (van Eemeren 2009).

to investigate how migration to the web has influenced the textuality of argumentative discourse, with a specific focus on cohesion. In order to define the object of analysis more closely, since argumentative discourse is too wide and heterogeneous a field, the study focuses on NGOs' campaigns, and in particular on communication with the public at large. As mentioned above the choice fell on this specific genre for two reasons: first, campaigns are overtly argumentative, relying on an appeal to rationality in order to win the audience's consensus and persuade them to take the desired action; second, this genre has known an unprecedented expansion thanks to the web, both in terms of visibility and of the amount of material that is made available to the public at large. Having generally been excluded from mainstream media in the pre-digital era, campaigns had to rely on word-of-mouth and leaflets/brochures that were distributed manually. On the web, space restrictions imposed by the necessarily limited size of the printed leaflet's page disappear, and the hypertext format in particular allows a much broader coverage of the issues and the initiatives related to a certain campaign. However, if on the one hand this undoubtedly represents an advantage, on the other hand, the loss of page boundaries (Storrer 2002) and the exponential increase of material might disorient the readers/users, and weaken the argumentative force of the message, with the much dreaded risk of failing to win their support to the campaign.

The analysis is carried out on two representative campaigns, both related to the food industry: Baby Milk Action, aimed against malpractices of infant formula marketing, and targeting Nestlé in particular for its role as market leader in this industry; and Greenpeace's campaign against genetically modified food. From a methodological point of view, the hypertexts taken into examination will be analysed, with special regard for cohesive devices, bringing in insights from the different disciplines and strands of research delineated in § 2. The higher levels of Swalesian categorization (communicative purpose and moves/links), accounting for the macro-organization of discourse, will be put in relation with the stages of the critical discussion posited by the pragmadialectical theory, so as to reconstruct the structure of the argumentation stretching throughout the hypertext, while also relying on the categories and methodological tools of multimodality (cf. Garzone *et al.* 2007) and CDA.

4. Analysis

While discussing coherence in argumentative hypertexts, I will draw on the distinction between local and global coherence (cf. Storrer 2002: 4), with the former referring to the logical relation between two "neighbouring parts of a text" and the latter to the "linkage of text constituents as it is mediated by the global theme addressed in the document, as well as by its rhetorical function in a larger context" (Storrer 2002: 4). Relying on these categories, Stutterheim (1997: 31) distinguished between *dynamic coherence*, which concerns the different 'information chunks' that build up into a text, and static coherence, which refers to the more general mental frame/script against which the single information chunks are interpreted.

In the case of NGOs' campaigns, in the digital era as well as before, the global frame of reference corresponds to the background knowledge the receiver has of the genre, and especially of its communicative purpose, i.e. winning new supporters for the campaign itself. It can therefore be expected that all the single moves which characterise the genre will be interpreted in light of this general purpose.

As regards the rhetorical moves, in the absence – to my knowledge – of a systematized codification for the genre of campaigns' homepage (as well as of its printed counterpart), I propose the following tentative classification:

Core moves:	Stating the campaign's aim Making the case for support Call to action
Optional moves:	Setting premises Establishing the NGO's legitimacy Contesting the legitimacy of the opponent Pre-empting possible counterclaims/objections.

On the ground of campaigns' inherently argumentative nature, a relation can be established between the moves – the core ones at least –

and the stages of the critical discussion model proposed by the pragmadialectic school. In this respect, 'stating the aim' could roughly correspond to the confrontation stage, in which a difference of opinion emerges. 'Making the case' would be equivalent to the argumentation stage, in which arguments are put forth in defence of one's standpoint and/or against the antagonist's standpoint; and the call to action could be seen as a conclusion in which the protagonist restates the validity of his/her positions. The following leaflet will illustrate this point:

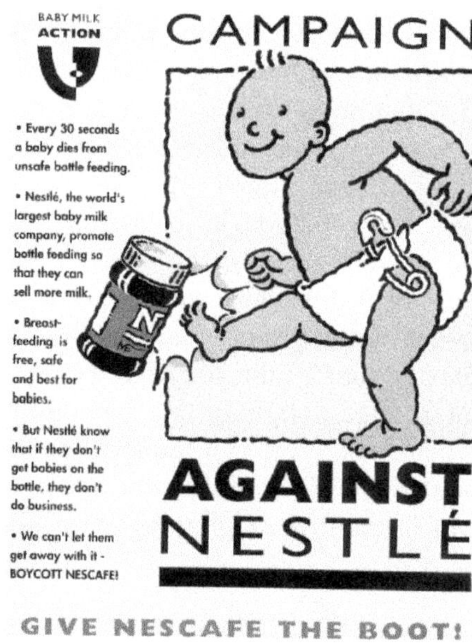

Figure 1. Baby Milk Action leaflet.

Stating the aim. As clearly stated in the leaflet, the aim of the campaign is to attack Nestlé. From the perspective of a critical discussion, this intention reveals a difference of opinion between the campaigners, who think that Nestlé should be opposed and the company in the first place, but possibly also any consumer raising doubts as to the necessity of a boycott. As is often the case in argumentation this move encompasses both the confrontation stage and the opening stage, in which the roles of protagonist (campaigners) and antagonist (Nestlé) are distributed.

Making the case. The standpoint that Nestlé should be opposed is defended by putting forth a series of arguments: Nestlé promotes bottle feeding, which in unsafe hygienic conditions causes high rates of mortality, to the detriment of breastfeeding which is safer and healthier.

Call to action. Following from what has been argued so far it is concluded that it is morally right to boycott Nestlé, therefore 'you' should boycott Nescafe, i.e. 'give Nescafe the boot!'
The critical discussion thus reconstructed can be represented as follows:

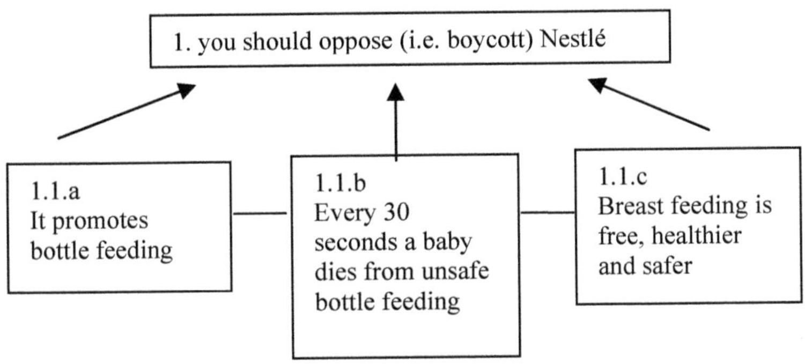

Figure 2. Baby Milk Action: reconstruction of critical discussion in the leaflet.

The process of cohesion building is here enhanced by graphic space constraints, which encourage the receiver to interpret all the textual and visual chunks contained within the page as mutually relevant, against the background of the general mental frame associated with NGOs' campaigns, thus determining global coherence. Similarly, local coherence is mainly constructed through the use of a graphic device, i.e. the bullet list, which links the different arguments, signalling that they are all parts of a superordinate textual unit.

Layout devices and space organization can then be seen as building a scaffold that guides the receiver in the correct interpretation of the message, and the scaffold seems solid enough as to pre-empt the need for linguistic connectors to make the relation between text constituents explicit.

4.1. *A failed attempt at creating coherence at hypertextual level*

Let us turn now to hypertext, and see what happens to textual coherence. The following screenshot, taken from Babymilk Action's page dedicated to the Nestlé Boycott, follows an overview of the initiatives against the Swiss giant, and is the first explicitly argumentative unit of text to be found on the page, as made clear by the title 'Evidence of Nestlé's malpractice':

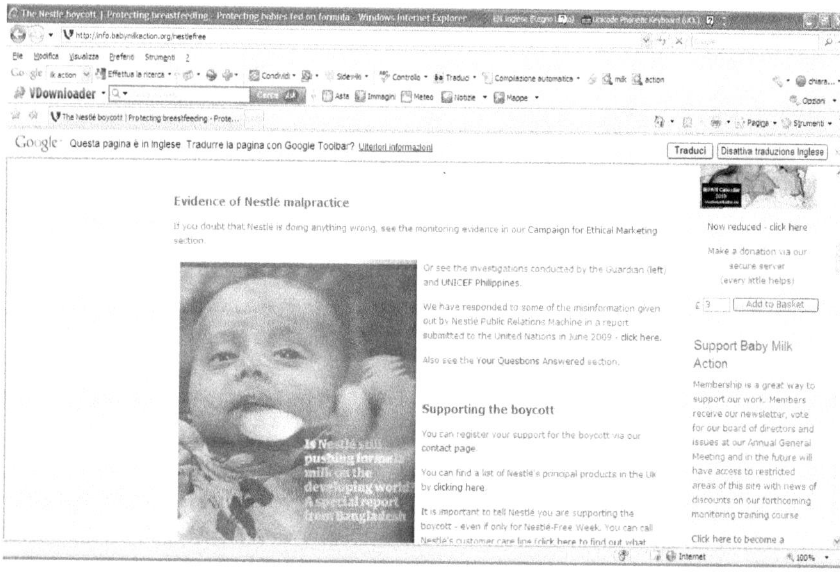

Figure 3. Screenshot of the Baby Milk Action website.

Before proceeding to analyse the structure of argumentation, a brief remark on the wording of the headline itself is in order. The act of providing evidence in support of an allegation activates the frame of a pseudo-legal discourse, which on the one hand implicitly constructs Nestlé's behaviour as a crime, rather than a socially irresponsible marketing strategy, and on the other hand presents what follows as part of an overt attempt to convince the reader by making facts available on the basis of which s/he will be able to form an independent opinion. However, the headline is worded in such a way as to increase the factuality of the allegations levelled at Nestlé, thus covertly orienting the reader to adhere to the view that Nestlé is responsible for 'malpractice'. To this end, the arguments put forth are presented as evidence, i.e. as something that will uncontroversially testify to Nestlé's guiltiness, while at the same time the information structure represents 'Nestlé's malpractice' as given – and therefore uncontroversial – information. If the same notion were presented as a process (e.g. "We think that Nestlé is guilty of malpractice and we will give you evidence for this") the attention of the reader would be drawn to the judgement expressed "Nestlé is guilty of malpractice" (new information), thus risking expo-sure it to a more critical scrutiny by the reader.

Resuming the analysis of the argumentative structure, the headline functions as a standpoint that is supported by a series of arguments, access to which is provided by the links embedded in the short units of text displayed under the headline. In turn, the standpoint and the arguments can be seen as supporting the more general claim, unexpressed at this level of the hypertext, "you should boycott Nestlé". The different levels of the hypertext correspond then to different levels of a subordinate argumentation structure that can be reconstructed as follows:

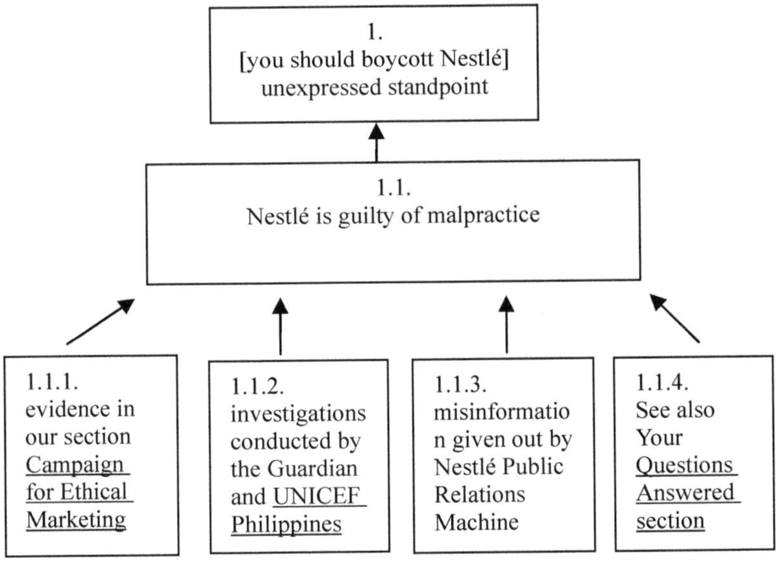

Figure 4. Baby Milk Action: reconstruction of website critical discussion.

Following the first line of defence (1.1.1.), the link gives access to these contents:

> **According to UNICEF: "Improved breastfeeding practices and reduction of artificial feeding could save an estimated 1.5 million children a year." Yet baby food companies continue to market artificial foods in ways that undermine breastfeeding. The people responsible have names and addresses – call on them to market products ethically.**
>
> Join Baby Milk Action to support the campaign and receive the Update newsletter.
>
> Nestlé is found to be the worst of the companies, so it is also targeted by an international boycott. For background and action, see the Nestlé-Free Zone.
>
> Please report instances of unethical marketing of baby foods in your area. Take action to make a difference. Also see the international monitoring report Breaking the Rules, Stretching the Rules 2007, later reports from the International Code Documentation Centre and some examples from Europe presented at the World Health Assembly in May 2006.
>
> We are finding it very difficult to keep up with demand for targeting violations due to staffing shortages. Can you help to keep the campaign going by becoming a member or sending a donation via our on-line Virtual Shop?

The first lexia proposes the same argument as we have illustrated with regard to the brochure (Baby food companies promote infant formula, thus undermining breastfeeding), followed by a call to action ("The people responsible have names and addresses – call on them to market products ethically"). The receiver can then interpret this text constituent as coherently related to the argumentative discourse started on the homepage. The lexias that follow, however, do not find a place in such a discussion, as they do not set forth any further evidence of Nestlé's malpractice, as might have been expected, but invite the user to take different actions (joining Baby Milk Action, finding out more about the boycott, reporting instances of unethical marketing of baby foods…). In this way a text announced as argumentative (evidence of malpractice) turns into instructive (Werlich 1983: 121), without any cues (Storrer 2002) that help the reader make sense of this shift. Scrolling down the page a huge table follows with links to 'action sheets' spanning over twelve years, which offer such a wealth of material as to make it impossible for the reader to keep track of global coherence.

The second link leads to a long reportage on the effects of bottle-feeding in developing countries (announced as the Philippines in the link, but actually turning out to be Bangladesh) functioning as a second line of defence (1.1.2). The beginning is reported here below:

> **Milking it**
> It was in 1977 that campaigners first called for a boycott of Nestlé because of its aggressive marketing of formula milk in the developing world. Thirty years on, have Nestlé and the other baby-milk firms cleaned up their act? Joanna Moorhead travels to Bangladesh to find out.
> Eti Khuman's face lies cradled on her mother's shoulder, her cheek resting in against Mina's collarbone. Eti is beautiful, but she is poorly: her breathing is heavy, and Mina has the distracted look of a mother who is very worried indeed. Eti's illness – first vomiting, then diarrhea – struck without warning. Like all mothers in Bangladesh, Mina knew to fear diarrhoea: in this country, diarrhoea can kill. So she wasted no time in bringing her eight-week-old daughter here, to the main diarrhoea hospital near her home in the capital, Dhaka.

After the first paragraph, which invites the reader to pass a judgment on Nestlé's conduct on the basis of subsequent information, the text steers abruptly from argumentation to subjective narration (Werlich 1983: 55). The focus, widening from the child's face cradled on her mother's shoulder to the scourge of diarrhea in Bangladesh, testifies to the intent of personalizing the issue, which had so far been dealt with in more abstract terms. Relying on Perelman and Olbrects-Tyteca's New Rhetoric (1969: 117) such a move can be interpreted as a way to increase the 'presence' of the data one selects for argumentation:

> One of the preoccupations of a speaker is to make present, by verbal magic alone, what is actually absent but what he considers important to his argument or, by making them more present, to enhance the value of some of the elements of which one has actually been made conscious.

A typical device used to enhance the 'presence' of an element is to have recourse to concrete objects such as, for example, Cesar's bloody tunic shown by Anthony to the Roman populace. The narrative concerning the poor conditions of just one child, who is thus made to symbolize the problem, seems to have precisely the function of – borrowing Bacon's words on the function of rhetoric – "applying reason to imagination, for the better moving of the will" (Perelman/Olbrechts-Tyteca 1969: 117).

Although in Perelman and Olbrechts-Tyteca's treaty presence is not associated with emotions, being defined as "a psychological datum operative already at the level of perception" (1969: 116), the examples of the bloody tunic and the child suggest that presence can be enhanced through pathos, which in argumentation has the purpose of appealing to emotions in the attempt to influence the audience's disposition to be persuaded.[4]

On the other hand, such a shift from the argumentative text type to subjective narration might jeopardize the global coherence of argumentative discourse, considering that the construction of coherent mental models is favoured by topic continuity and concurrence with conventional text patterns (Storrer 2002). However, in the light of the above considerations on 'presence' and the use of emotions for argumentative purposes, this possible disadvantage is amply compensated for by the effectiveness of this choice in supporting rational appeals. Besides, recourse to subjective narration can be seen as a way to diversify the lines of defense so as to cater for different audiences, some people being more easily moved through emotion than through reasoning. What seems to hopelessly disorient the reader, however, is the length of the text (equivalent to six pages of thick linear text) associated to the lack of compliance with web writing conventions, which altogether hinder the process of coherence building and – unlike the abrupt shift to the narrative mode – are hardly justifiable.

To sum up, the analysis so far conducted has shown how problematic arguments on the web can be, when the affordances of this environment are misused. Particularly, unjustified shifts from one text type to another in contiguous or otherwise related lexias, and an unrestrained expansion of texts due to the virtually unlimited 'space' available can disrupt the coherence and the effectiveness of arguments.

---

4 For the role of emotions in argumentation, see Plantin (1999, 2004) and Micheli (2010). For a more specific focus on rhetorical appeals – including appeals to pathos – in fundraising discourse, an activity akin to NGOs' campaigns, cf. Biber *et al.* 2007, chapter 5.

4.2. A successful case of hypertextual coherence

We shall now consider the second case study, i.e. Greenpeace's campaign against GMOs, whose homepage is reported below (cf. Figure 2). Comparing it with the Baby Milk Action's page, which displayed a headline and a list of links, the main difference is that in this case the reader finds a short statement making the case for the campaign:

> **Say no to GE**
> While scientific progress on molecular biology has a great potential to increase our understanding of nature and provide new medical tools, it should not be used as justification to turn the environment into a giant genetic experiment by commercial interests. The biodiversity and environmental integrity of the world's food supply is too important to our survival to be put at risk.
> **Your help is needed NOW!**

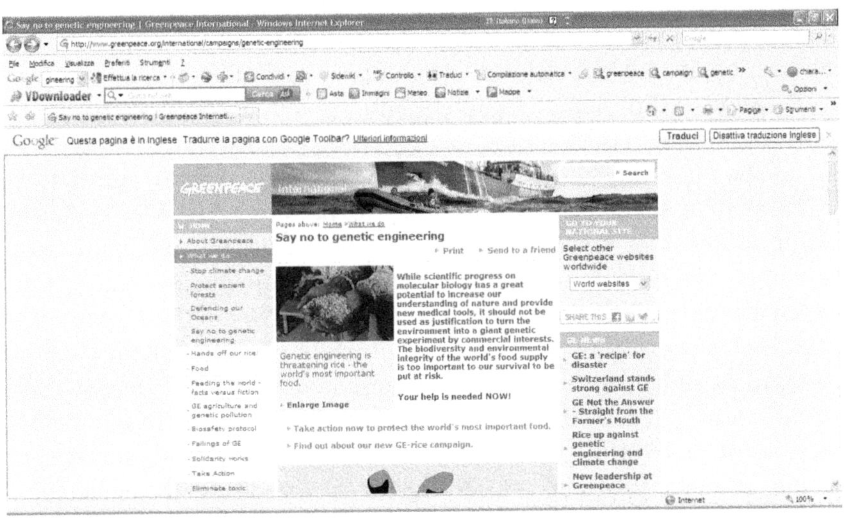

Figure 5. Screenshot of Greenpeace's web campaign "Say no to genetic engineering".

From a linguistic point of view, the text immediately classifies as a mixture of instructive and argumentative text types. In line with the purpose of all campaigns, i.e. to influence people's behaviour, the instructive component accomplishes the function of asking the reader to do something, namely to say no to genetic engineering while the argumentative component justifies such a request. In terms of Werlich's text grammar categories, the headline 'Say no to…' is an action-demanding sentence (here coming in the form of a command), which is the typical thematic text base of instructive texts (1983: 29).

The body of the text, instead, pivots on the standpoint "progress on molecular biology should not be used as justification to turn the environment into a giant genetic experiment by commercial interests", and corresponds to the cognitive function associated with the argumentative text type, i.e. judging in response to a problem. The argumentative nature of the text is confirmed also by elements of textual coherence. Looking at the logical relation linking the different constituents of this text, the first proposition ("While scientific progress on molecular biology has a great potential to increase our understanding of nature and provide new medical tools") performs the rhetorical function of making a concession from which an endorsement of GE would be expected to follow, only to frustrate such an expectation in the subsequent statement ("it should not be used as justification to turn the environment into a giant genetic experiment by commercial interests").

The use of concessive constructions is often attested in argumentative texts, where it is used to pre-empt possible criticism and/or establish a sort of common ground with the antagonist, by showing a "favourable reception to some of the opponent's real or presumed arguments" (Perelman/Olbrechts-Tyteca 1969: 488). Going back to Werlich's categories, the two propositions form an antithetical introduction, again a typical trait of argumentative text types. The 'neighbouring' proposition "The biodiversity and environmental integrity of the world's food supply is too important to our survival to be put at risk" functions as an argument supporting the standpoint "[scientific progress on molecular biology] should not be used as justification […] interests". Finally, the peroration "Your help is needed now" circularly closes the paragraph by reverting to the instructive text type, and thus creates a further pattern of coherence with the headline.

The first lexia is followed by two links, which exhort the user to take some form of action ("Take action now to protect the world's most important food" and "Find out about our new GE-rice campaign"), while scrolling down the page two other lexias appear:

What's wrong with genetic engineering (GE)?

expository

Genetic engineering enables scientists to create plants, animals and micro-organisms by manipulating genes in a way that does not occur naturally.

These genetically modified organisms (GMOs) can spread through nature and interbreed with natural organisms, thereby contaminating non-'GE' environments and future generations in an unforeseeable and uncontrollable way.

argumentative

Their release is 'genetic pollution' and is a major threat because GMOs cannot be recalled once released into the environment.

Because of commercial interests, the public is being denied the right to know about GE ingredients in the food chain, and therefore losing the right to avoid them despite the presence of labelling laws in certain countries.

[…]

We believe:

GMOs should not be released into the environment since there is not an adequate scientific understanding of their impact on the environment and human health.

We advocate immediate interim measures such as labelling of GE ingredients, and the segregation of genetically engineered crops and seeds from conventional ones.

We also oppose all patents on plants, animals and humans, as well as patents on their genes. Life is not an industrial commodity. When we force life forms and our world's food supply to conform to human economic models rather than their natural ones, we do so at our own peril.

Although its headline "what's wrong with GE" presents it as argumentative, the first of these two lexias starts out as an expository text, explaining what GE is with recourse to definitions. Its function is that of bridging a possible gap between the writer's and the receiver's knowledge and in this sense it represents an example of the peripheral moves of NGOs' campaigns that I have labeled 'Setting premises'. However, it is not an objective and neutral explanation, as the use of highly evaluative language (cf. the words and expressions underlined) clearly tries to cast a negative light on GE. Once the premises have been set, the text takes an openly argumentative tone, made even more explicit by the declarative headline "We believe".

The analysis conducted so far shows that the home page of Greenpeace's campaign against GMOs presents strong textual coherence, both within and across single lexias. We will now consider lower levels in the hypertext hierarchy, by looking at the content made available through the first two links in the page:

> Rice is daily food for half of the world's population. Genetically modified (GM) rice, on the other hand, is a threat to our agriculture, our biodiversity and a possible risk to our health.
> At present, GM rice is not grown commercially anywhere in the world. But Bayer, the German chemical giant, has genetically manipulated rice to withstand higher doses of a toxic pesticide called glufosinate, which is considered to be so dangerous to humans and the environment that it will soon be banned from Europe. (Link 'Take action now to protect the world's most important food')

Going a step down in the hypertext hierarchy, we can see that the focus narrows from the general topic of GMOs, to genetically modified rice, a shift anticipated on the homepage, which features a picture of two hands holding rice, with the caption "Genetic engineering is threatening rice – the world's most important food". As a further cohesive device, the latter expression (the world's most important food), which in the caption is an apposition, is repeated also in the link, where it is used as a substitute for the word 'rice'.

Although the exhortation to take action contained in the link suggests that it will give access to instructions for joining the campaign, the content of the node is in fact argumentative. In the first place the premises of the campaign are made explicit, with a view to building a common ground between the sender and the receiver: by presenting rice as the staple food for half the world's population, the writer prepares the ground for persuading readers of the urgency of the alarm conveyed by the standpoint: "Genetically modified (GM) rice, on the other hand, is a threat to our agriculture, our biodiversity and a possible risk to our health". The standpoint is defended by putting forth the argument that Bayer has genetically modified rice to resist higher doses of a toxic pesticide that the EU has banned, hence the reason for alarm. The node accessed through the second link, (*Find out about our new GM campaign*) repeats these concepts and expands on them:

> **Hands off our rice!**

Figure 6. Detail of Greenpeace's webpage.

> **The big issue**
> Rice is the world's most important staple food – with more than half of the global population eating it every day. It has been grown around the world for over 10,000 years and is cultivated in 113 countries. Rice is also a key ingredient in a wide variety of processed foods ranging from baby food to the more obvious rice noodles. *But all this is under threat as genetic engineering (GE) continues to creep up on our most valuable food.*
>
> Today, genetically modified (GM) rice only exists in field trials. But all that could change tomorrow as agri-chemical companies and some governments around the globe are trying to commercialise it. Ecological farming is the safest solution to the food crisis and looming climate change disasters. Keeping rice GE-free is not just about consumer choice or the environment – it's a lot bigger than that. It's a matter of global food security, human rights and survival.

As can be noticed, the structure identified in the previous node is repeated here: an expository introduction makes the premises explicit, thus paving the way for the enunciation of the standpoint (*But all this is under threat as genetic engineering (GE) continues to creep up on our most valuable food*), which is then followed by a concession aimed to pre-empt the possible objection that GM rice is not a threat as it is not commercialised. The critical discussion closes with a final peroration that, in a rhetorical crescendo, justifies the superiority of green farming over GM not only in terms of nutrition safety, but also of human rights and survival.

Finally, this last node prompts a brief remark on the use of headlines. In newspapers headlines have the function of attracting the attention of the readers, with a view to inducing them to read the story. In a similar fashion, this website reveals great attention to the construction of headlines, which include puns and intertextual references. In the example above, 'Hands off our rice', the noun group 'our rice' seems to echo the evocative 'our bread' of the *Holy Father*, not as an appeal to religious authority, but rather as an established topos in Western culture, which cannot fail to sound as an indirect assertion of the inalienability of the right to food.

At first sight this interpretation of the headline may seem far-fetched. However, a clue to the intentionality of such an intertextual reference comes from other headlines featuring on the same page, in which this mechanism is definitely more explicit, as in the case of the exhortation "Stand up for your rice!" Here the reader will easily seize the reference to Bob Marley's popular line "Stand up for your rights", with the relation made even stronger by the assonance between rice /raɪs/ and /raɪts/. Also in this case the reference not only has the function of catching the reader's attention, but also indirectly activates an inference, i.e. that food is a right.

The analysis of the Greenpeace case study shows that a coherent argumentative discourse can be produced in computer-mediated communication in spite of the lack of linearity inherent in hypertexts. Relying on primacy effects, the writer can hope to impose a certain pattern of reading. The first lexia encountered on the home page thus starts making the case for opposition to GE in general terms, introducing through peripheral lexias the specific campaign for GM-free rice. Lower-level nodes are to be interpreted in the light of this general frame, but they are constructed in such a way as to make them self-standing, so that even if the reader fails to keep track of their relation with overarching nodes, s/he will still find a coherent and complete argumentative text unit contained in each single node. Exploiting recency effects, nodes tend to close with exhortative remarks to take action or with a forceful reassertion of the point made within them, in the hope that this, at least, will be the bring-home message. Repetition effects, namely the recurrence of symmetrical structures of argumentation in different nodes, finally increase the chances that, even adopting a selective-reading mode, the reader will encounter those messages that are considered core.

5. Conclusion

The analysis of two cases of online NGOs' campaigns has provided a preliminary understanding of how the influence of the medium affects the textuality of argumentative discourse, particularly in terms of coherence design in hypertexts. With reference to Swales' three-level model of genre, revisited by Askehave and Ellerup Nielsen's to suit the analysis of digital genres, special attention has been devoted to the level of moves, for two reasons: 1. by creating conventional text patterns mutually recognized by the reader and the writer, they contribute to the construction of coherence; 2. it is the level that is most affected by the migration of genres onto the web, as hypertext does not allow for a linear organization of the text into moves.

On the basis of the cases analysed, some conclusions can be drawn as to the functioning of coherence in argumentative hypertexts. The analysis of Baby Milk Action's page dedicated to the Nestlé boycott – where the author seems to have abdicated his/her responsibility for selecting and arranging contents into a coherent global structure – demonstrates that the risk of weakening coherence relations exists due to the potentially dispersive character of hypertexts. In the Baby Milk campaign coherence is granted only in terms of topic continuity, but it is almost impossible to recognize a conventional text pattern, due mainly to the excess of information made available on the website and the lack of an adequate textual scaffolding, all to the detriment of the effectiveness of argumentation in persuading the reader. The Greenpeace campaign, on the contrary, has been successful in imposing on the hypertext what Storrer (2002) calls 'dynamic coherence planning'. Two types of strategies seem to have been deployed to this end: on the one hand primacy effects, alongside headlines aimed at grabbing attention, try to suggest a pattern of reading which reflects a hierarchical organization of issues, thus contributing to the creation of global coherence. In Greenpeace's campaign the case for rejecting GE is made in very general terms on the home page, while lower levels, made available through links, narrow the focus on specific campaigns (e.g. the one for GM-free rice). On the other hand, argumentation presents a 'modular' structure, where each node contains a self-standing critical discussion, which in turn generally features all the core moves of the genre (expressing the standpoint, supporting it through arguments, calling to action). While global coherence is taken care of by the attempt to suggest a pattern of reading, this strategy enhances local coherence. Furthermore, the same critical discussion, with slight variations, is repeated in more than one node, so that different navigation patterns lead to the same message.

To conclude, even though hypertexts do not allow for linear author-imposed textual coherence, which would require a sequential reading mode, this study has shown that it is still possible to design argumentatively effective texts on the Web. The only necessary caution is for authors to take into account the modalities through which texts will be accessed. The virtually unlimited availability of space, far from inducing authors to flood the reader with a mass of information, should call for a rigorous selection of arguments. Furthermore, rather than conceiving each level as part of a single critical discussion, leaving to the reader the responsibility for keeping trace of its global coherence, it seems preferable to design short self-standing discussions, repeated with slight variations, so as to be accessible from different reading paths.

The findings of this chapter would benefit from an extension of the analysis to other websites characterized by an argumentative drive, which might reveal other challenges posed by the hypertext environment as well as other strategies for enhancing cohesion in arguments on the web. At the same time, a qualitative analysis of NGOs' websites conducted on a larger scale could lead to a better definition of the genre 'NGO's campaign', which is to the best of my knowledge, quite an under-explored genre in institutional communication.

References

Aakhus, Mark / Jackson, Sally 2005. Technology, Interaction, and Design. In Fitch, Kristine L. / Sanders, Robert E. (eds) *Handbook of Language and Social Interaction.* Mahwah, NJ: Lawrence Erlbaum, 411-436.

Askehave, Inger / Ellerup Nielsen, Anne 2005. Digital genres: a challenge to traditional genre theory. *IT & People* 18/2, 120-141.

Bekalu, Mesfin A. 2006, Presupposition in News Discourse. *Discourse and Society.* 17/2, 147-172.

Bernstein, Mark 1998. Patterns of Hypertext. *Proceedings of Hypertext '98.* Pittsburgh: Association of Computing Machinery, 21-29.

Bhatia, Vijay K. 1993. *Analysing Genre- Language Use in Professional Settings.* London: Longman.

Biber, Douglas / Connor, Ulla / Upton, Thomas 2007. *Discourse on the Move Using corpus analysis to describe discourse structure.* Amsterdam: John Benjamins.

Bolter, Jay D. 1991. Topographic writing: Hypertext and the electronic writing space. In Delany, Paul / Landow, George P. (eds) *Hypermedia and literary studies.* Cambridge MA: MIT, 105-118.

Bolter, Jay D. 2001. *Writing Space: The Computer, Hypertext, and the History of Writing.* Hillsdale, New Jersey: Lawrence Erlbaum (second edition).

Carter, Locke M. 2000. Arguments in Hypertext: A Rhetorical Approach. *Proceedings of the Eleventh ACM on Hypertext and Hypermedia*, TX.

Carter, Locke M. 2003. Argument in hypertext: Writing strategies and the problem of order in a nonsequential world. *Computers and Composition.* 20, 3-22.

Crowston, Kevin / Williams, Marie 1997. Reproduced and Emergent Genres of Communication on the World-Wide Web. *Proceedings of the 30th Hawaii International Conference on System Sciences: Digital Documents - Volume 6*.

De Beaugrande, Robert-Alain / Dressler, Wolfgang U. 1981. *An Introduction to Text Linguistics.* London: Longman.

Finnemann, Niels O. 1999. Hypertext and the Representational Capacities of the Binary Alphabet. Available at <www.hum.au.dk/ckulturf/pages/publications/nof/hypertext.htm>.

Fritz, Gerd 1998. Coherence in Hypertext. In Wolfram, Bublitz / Lenk, Uta / Ventola, Eija (eds) *Coherence in Spoken and Written Discourse: How to Create It and How to Describe It.* John Benjamins: Amsterdam.

Garzone, Giuliana 1997. Osservazioni sul valore pedagogico dell'ipertesto. In Schena, Leo / Garzone, Giuliana (eds) *Il computer al servizio degli insegnamenti linguistici, Milano 28.4.1995-3.5.1996.* Milan: Centro Linguistico Università Bocconi, 21-39.

Garzone, Giuliana 2007. Genres, multimodality and the World-Wide Web: Theoretical Issues. In Garzone *et al.*, 15-30.

Garzone, Giuliana E. / Catenaccio, Paola / Poncini, Gina (eds) 2007. *Multimodality in Corporate Communication. Web Genres and Discursive Identity.* Milano: FrancoAngeli.

Hoffman, Donna L. / Novak, Thomas P. 1996. Marketing in Hypermedia Computer-Mediated Environments: Conceptual Foundations. *Journal of Marketing.* 60/3, 50-68.

Houtlosser, Peter. 2006. Strategic Manoeuvring: A Synthetic Recapitulation. *Argumentation* 20, 381-392.

Hovland, Carl I. 1957. *The Order of Presentation in Persuasion.* New Haven: Yale UP.

Jackson, Sally. 1998. Disputation by design. *Argumentation.* 12/2,: 183-198.

Johnson, Ralph H. / Blair, Anthony J. 1978. Introduction. In Johnson, Ralph H. / Blair, Antony J. (eds) *Informal Logic: First International Symposium.* Inverness, Ca: Edge, 1-25.

Kolb, David 1997. *Scholarly Hypertext: Self-Represented Complexity. Proceedings of Hypertext '97.* Southhampton: Association of Computing Machinery, 247-268.

Kress, Gunther / van Leeuwen, Theo 2001. *Multimodal Discourse: The Modes and Media of Contemporary Communication.* Arnold: London.

Landow, George P. 1992. *Hypertext: the Convergence of Contemporary Critical Theory and Technology.* Baltimore: Johns Hopkins.

Lemke, Jay L. 2006. Towards Critical Multimedia Literacy: Technology, Research, and Politics. In Reinking D. / Labbo L. / McKenna M. / Kiefer R. (eds) *International Handbook of Literacy & Technology, v2.0*. Hillsdale NJ: Lawrence Erlbaum Associates.

Lewiński, Marcin. 2010. *Internet political discussion forums as an argumentative activity type. A pragma-dialectical analysis of online forms of strategic manoeuvring in reacting critically*. Amsterdam: SicSat. Available at: <http://dare.uva.nl/en/record/340631>.

Micheli, Raphaël 2010. Emotions as Objects of Argumentative Constructions. *Argumentation*. 24/1, 1-17.

Perelman, Chaim / Olbrects-Tyteca, Lucie 1969. *The New Rhetoric. A Treatise on Argumentation*. Paris: Notre Dame.

Plantin, Christian 1999. Arguing emotions. In van Eemeren, Frans *et al.* (eds) *Proceedings of the Fourth International Conference of the International Society for the Study of Argumentation*. Amsterdam: SicSat, 631-638.

Plantin, Christian 2004. On the Inseparability of Reason and Emotion in Argumentation. In Weigand, Edda (ed.) *Emotion in Dialogic Interaction*. London: John Benjamins, 269-281.

Prakken, Henry 2005. AI and law, logic and argumentation schemes. *Argumentation*. 19, 303-320.

Storrer, Angelika 2002. Coherence in Text and Hypertext. (Preprint). Then published in *Document Design* 2002. Available at <http://www.hytext.info>.

Stutterheim, Christiane von. 1997. *Einige Prinzipien des Textaufbaus. Empirische Untersuchungen zur Produktion mündlicher Texte*. Tübingen: Niemeyer.

Swales, John 1990. *Genre Analysis: English in Academic and Research Settings*. Cambridge: Cambridge University Press.

Tosca, Susana P. 2000. A Pragmatics of Links. *Journal of Digital Information*. 1/6. Available at <http://journals.tdl.org/jodi/article/viewArticle/23/24>.

Toulmin, Stephen E. 1958/2003. *The Uses of Argument*. Cambridge: Cambridge University Press.

van Dijk, Teun A. 1988. *News Analysis*. Hillsdale and London: Lawrence Erlbaum Associates.

van Dijk, Teun A. 2006. Discourse and Manipulation. *Discourse and Society.* 17/3: 359-383.

van Eemeren, Frans H. (ed.) 2009. *Examining Argumentation in Context: Fifteen Studies on Strategic Maneuvering.* Amsterdam/ Philadelphia: John Benjamins.

van Eemeren, Frans H. 2010. *Strategic Manoeuvring in Argumentative Discourse: Extending the Pragma-dialectical Theory of Argumentation.* Amsterdam: John Benjamins.

van Eemeren, Frans H. / Grootendorst, Rob. / Jackson, Sally / Jacobs, Scott 1993. *Reconstructing Argumentative Discourse. Studies in Rhetoric and Communication.* Tuscaloosa: The University of Alabama Press.

van Eemeren, Frans H. / Grootendorst, Rob / Snoeck Henkemans, Francisca. 1996. *Fundamentals of Argumentation Theory: A Handbook of Historical Background and Contemporary Developments.* Mahwah, N.J., U.S.A.: Erlbaum.

Walton, Douglas / Reed, Christian / Macagno, Fabrizio 2008. *Argumentation Schemes.* Cambridge: Cambridge University Press.

Weger, Jr., Harry / Aakhus, Mark 2003. Arguing in Internet chat rooms: Argumentative adaptations to chat room design and some consequences for public deliberation at a distance. *Argumentation and Advocacy.* 40/1, 23-38.

Werlich, Egon 1983. *A Text Grammar of English.* Heidelberg: Quelle und Meyer.

Wodak, Ruth / Meyer, Michael (eds) 2001. *Methods of Critical Discourse Analysis.* London: Sage.

Web 2.0

ELIZABETH ROWLEY-JOLIVET

Open Science and the Re-purposing of Genre: An Analysis of Web-mediated Laboratory Protocols

1. Introduction

The Internet, and in particular the advent of Web 2.0, has transformed the way in which many aspects of science are conducted and communicated. By providing powerful, accessible and widely distributed computer resources, the Internet has enabled scientific and technological collaboration on a hitherto unprecedented scale: as David (2004) points out, the access to and sharing of digital research data and of information tools that facilitate the storage, search, retrieval, and analysis of the data, have "stimulated the emergence of entirely new forms of distributed research collaboration and information production" (David 2004: 6). These transformations, variously referred to as cyberscience (Nentwich 2003), e-science (Jankowski 2007), or Science 2.0 (Waldrop 2008), have generated several different kinds of collaborative initiatives. Many of these developments, though new from a technological point of view, can be seen as modern, Internet-enabled embodiments of the Mertonian (Merton 1973) ideals of science: communalism, or the common ownership of scientific discoveries; universalism, according to which claims to truth are evaluated in terms of universal or impersonal criteria; disinterestedness, according to which scientists are rewarded for acting in ways that outwardly appear to be selfless; and organized skepticism – all ideas are subjected to rigorous, structured community scrutiny. As Waldrop (2008) comments, "for Science 2.0 advocates, the real significance of Web technologies is their potential to move researchers away from an obsessive focus on priority and publication, toward the kind of openness and community that were supposed to be the hallmark of science in the first place".

1.1. Open Access vs. Open Science

One of the earliest of these collaborative initiatives was open access journals, i.e. journals that do not charge readers or their institutions for access, but allow users to read, download, and search the full texts of research articles. The pioneer in this field was the preprint server arXiv launched in 1991 in high-energy physics, which now accepts papers beyond this initial focus (see <http://arxiv.org>). It has since been followed by similar online libraries in other fields, such as Bio-Med Central (founded in 2000; 212 journals currently available) and the Public Library of Science (founded in 2003; 7 PLoS journals). The cross-disciplinary Directory of Open Access Journals currently lists 6,080 journals, all of which are peer-reviewed or ensure editorial quality control, on its website (<http://www.doaj.org>), confirming the rapid development and scale of open access publishing over the past decade. The common agenda of non-profit open access journals is the desire to break the stranglehold that paying subscription journals had acquired over scientific publication, and which made rapid, cost-free access to research results beyond the reach of many laboratories and researchers, thus limiting the circulation of knowledge and the development of research itself. Many major universities and labs have also created open repositories, where researchers can self-archive by uploading their publications: ROAR, the Registry of Open Access Repositories, currently lists 2,104 research repositories freely available online (<http://roar.eprints.org>). While concern has been voiced about issues such as intellectual property rights, priority claims, the quality of open access articles, and privacy, practical solutions such as Creative Commons licenses and time-stamping have already been found to many of these issues (see e.g. Borgman 2007).

A second important open access initiative enabled by Web 2.0 concerns the creation of vast online databases of scientific information, free of charge. This is an extremely active field, as it enables researchers to carry out comprehensive and rapid searches through an exponentially growing quantity of data. While such initiatives involve the sharing of information rather than of knowledge, these powerful search tools make a very significant contribution towards furthering scientific knowledge. A few recent examples in chemistry, where searches can be made using natural language, chemical formulae, or visual representations (2D and 3D molecule structures) are: emolecules, launched in 2005, a search engine for chemical molecules (<http://www.emolecules.com>); PubChem (<http://pubchem.ncbi.nlm.nih.gov/>) and ChemSpider (<http://www.chemspider.com>), freely accessible databases of millions of chemical compounds.

Taking these developments to their logical extreme, a further initiative that has recently emerged, and which is the focus of this chapter, is the Open Science movement. The main, and highly important, difference with the previous two initiatives is that open science data, like Wikipedia, are mutable: the documents evolve through a large number of edits made by a worldwide community of scientific contributors, and are collectively and gradually built up. The first tools were direct offshoots from the open source software model, and proposed the collaborative writing and free release of software dedicated to scientific applications (see <http://www.openscience.org>). Other initiatives created in the last five or six years focus on specific fields or other features of scientific activity. Open Notebook Science, for example, a term coined by Jean-Claude Bradley of Drexel University in a now-famous blog post on September 26th 2006, involves placing the personal laboratory notebook of the researcher online along with all raw and processed data and

> with the researcher's interpretation in a format that anyone can easily re-analyze, re-interpret and re-purpose. A good example of re-purposing is using some results and observations from a failed experiment in a way that was never intended by the original researcher. This just doesn't happen regularly in science because failed experiments are almost never included in publications (Bradley 2006).

These unpublished, or so-called 'dark data', give revealing insights into scientific activity in the lab, as will be shown in section 4.2. Along with chemistry, the biomedical field was also an early adopter of open science, one of the motivations being that the commercial interests of pharmaceutical firms had led to the neglect of areas of research that were deemed unprofitable (Kepler *et al.* 2006). The Synaptic Leap, launched in 2006, a collaborative research project into neglected tropical diseases, declares in its mission statement:

> We believe that effective collaborative community tools that promote transparent research and connect resources will: 1. generate ideas more quickly; 2. reduce redundant work. Thus, ultimately cures will be developed more quickly and more effectively. (<http://www.thesynapticleap.org>)

The open science initiative that will be analysed here is OpenWetWare, launched in 2005 by graduate students in biological engineering at MIT, initially in order to share and improve their lab protocols when trying to grow cell cultures or manipulate DNA. The mission statement of OWW reads as follows:

> The goals of OWW are to support open sharing of research, education, publication, and discussion in biological sciences and engineering. We promote and support collaborations among researchers, students, and others who are working towards these goals. We believe that open sharing of research improves the quality and pace of scientific and engineering research. (<http://www.openwetware.org>)

Since its launch, the OpenWetWare (henceforth OWW) site has burgeoned, and currently (January 2011) contains 17,950 content pages (i.e. excluding 'talk' pages, pages about OWW, minimal stub pages, redirects, etc.) contributed by 8,720 registered users. In addition to protocols, the site has an extensive databank of biological materials and other research resources (reference materials, organisms, links, help), hosts a blog and provides links to about 250 labs and research groups as well as information about some academic courses. The site is a wiki, and the protocols are collaboratively written and edited by the users.

1.2. Web-mediated genres

Many of the document types that have migrated to the Web are – like the OWW protocols – based on already existing and well-established discourse genres: books, research articles, encyclopedia entries, or company annual reports, for instance. As Orlikowski and Yates (1994) suggested, when faced with a new communicative situation, individuals will typically draw on their existing genre repertoire, reproducing genres they are already familiar with as members of other communities. Some of these genres have moved intact to the web, as .pdf downloads of existing documents, for example; others, such as the research article, though they have not undergone major changes in this migration process, have nevertheless accrued considerable 'added-value' (Luzón 2007), or new Internet-enabled features and functions, in the process. As several studies have pointed out (Garzone *et al.* 2007; Giltrow/Stein 2009; Myers 2010), the highly different affordances of Web 2.0 compared to traditional print media – multimodality, hypertextuality, non-linearity, collaborative writing and user-generated content, mutability – have induced significant changes in many existing genres, as they adapt to the new medium. In addition to changes in existing genres, new web-native genres appear to be emerging (Crowston/Williams 1996), entailing as yet unresolved implications for genre theory itself. The question that arises is whether traditional genre theory can handle these evolving web-mediated and novel web-native genres, or whether a 're-purposing' (Askehave/ Swales 2001) of genre theory is necessary (see Garzone 2007 and this volume for a fuller discussion of this ongoing debate). Askehave/Ellerup Nielsen (2005), for example, propose an 'up-grade' of the traditional genre model to incorporate media features that cannot be accounted for in existing theory, arguing that the medium itself should form part of a revised genre model, since "the medium adds unique properties to the web genre in terms of production, function and reception which cannot be ignored in the genre characterisation" (2005: 125).

As a contribution to this debate, the present chapter will focus on one long-standing scientific genre, laboratory protocols, and will compare a selection of protocols from the collaborative OpenWetWare website with traditional print protocols, in order to highlight their differences: what new properties does the Web-mediated genre possess, and to what extent do these new features modify or call into question the defining features of the original genre? The choice of the protocol genre stemmed from the following hypothesis: lab protocols are a rigidly formatted genre, exhibiting a minimal amount of variability from one print realization to another; it was therefore felt that any observable differences between the offline and online versions could be more reliably attributed to the web context than would be the case with a genre that possessed greater inherent flexibility.

The chapter is organized as follows. Section 2 gives a characterization of the traditional protocol genre, with which the web-mediated protocols can then be compared. After a brief presentation of the corpus (section 3), the comparative analysis in section 4 then focuses on three features that are widely considered to be central in genre characterization (Giltrow/Stein 2009): structure, communicative purpose, and linguistic form.

2. Characterisation of the lab protocol genre: structural and linguistic characteristics

Laboratory protocols are used in many fields of experimental science. They lay out the materials and procedure needed to perform the protocol, and are written by research labs, by authors of science manuals to instruct students and researchers, or by manufacturers to explain how to use their lab kits. A lab protocol is an instructional genre (Peikola *et al.* 2009; Murcia-Bielsa 2000), in which the communicative purpose is to tell the reader how to perform an action or series of actions. Both the structure and the language used are highly constrained, and can be briefly characterized as shown in Figure 1.

RNA FROM E. coli CELLS

Solutions:
1. Low phosphate medium (for 32P labeled cells): Use a rich broth such as 2 X TB. Add MgSO4 to 0.01 M then concentrated NH4OH to 1%. Let sit at least 2 hours and filter. Add Tris-HCl to 0.01 M and adjust to pH 7.4. Add 2 x 10-4 M phosphate.
2. Lysis buffer:
0.3 M Na Acetate pH 5.5
0.5% SDS
0.01 M EDTA
3. Redistilled or liquefied phenol equilibrated with sodium acetate pH 5.5 before heating.

Procedure:
1. For 32P RNA, grow overnight culture in low phosphate medium.
2. Dilute 1:200 with low-phosphate medium containing 2-5 mCi 32PO4 and grow at 37°C to a concentration of about 108 cells/ml (perhaps 5 hours).
3. Chase with one-third volume fresh medium containing 10-2 M phosphate for about 30 min.
4. Harvest cells by spinning 5 min at 5,000 RPM and resuspend the pellet in lysis buffer using 1 ml for every 10 ml culture medium.
5. Extract with hot phenol 10 min at 67°C using a Pasteur pipette for mixing.
6. Separate layers by centrifugation and re-extract the aqueous layer with hot phenol.
7. Add 2 vol. cold 100% ethanol to precipitate nucleic acids.
8. After collecting RNA by centrifugation, DNA may be removed by DNase treatment and/or salt washing.

Figure 1. Standard laboratory protocol.
Source: <http://groups.molbiosci.northwestern.edu/morimoto/research/Protocols>.

Structurally, lab protocols comprise an obligatory 'Procedural' section, presented, as in Figure 1, as a numbered list of steps to be carried out in chronological order. The length of this section will of course depend on the complexity of the procedure. This core section can, if relevant, be preceded by two types of optional sections: first, immediately after the title, a short paragraph, called the *Lead* here, giving the rationale for the protocol – its intended use, or what it is designed to do – or other relevant information. If the instructor considers that the audience is familiar with the purpose of the protocol, this introductory section is omitted. Far more common is the second type, which lists the materials required for the protocol: reagents, solutions, apparatus, etc., as in Figure1 above, which need to be prepared or set up before carrying out the procedure. In certain cases, other optional sections follow the core Procedure section: a comments or tips section, which either gives alternatives for some elements in the protocol (type of apparatus, quantity of reagents, etc.) or advice on how to solve commonly encountered problems; and lastly, if the protocol is based on a previous publication in a research article or manual, a literature reference to this source; this concluding section will be called the *Tail*. The structure can therefore be schematically represented as in Figure 2; optional elements are in italics; the order in which they appear, however, is fixed.

Name of Protocol

Lead

Materials
- *Reagents*
- *Equipment*

Procedure
Tail

Figure 2. Schematic structure of laboratory protocols.

Linguistically, lab protocols are highly conventionalized. In common with instructional texts in general, there are two speaker roles: the instructor, and the instructed (Peikola *et al.* 2009). The authority relationship between the interactants is unambiguous: the instructor, by virtue of his or her superior knowledge, adopts the position of authority, which is acknowledged by the reader. This authorizes the writer of the protocol to use strong deontic modality (*must*) and imperatives; while these are often felt to be face-threatening in other contexts (Brown/Levinson 1978), "instructions are far from being perceived as face-threatening orders. They are rather considered as advice or guides" (Murcia-Bielsa 2000: 128), and therefore require no mitigation or facework. As Figure 1 clearly shows, the imperative is the default syntactic structure in the protocol genre. Moreover, as Murcia-Bielsa further argues, syntactic expressions do not possess a fixed politeness level: their degree of politeness depends, in addition to the social roles of the interactants, on contextual features such as the mode (speech *vs.* writing) and on their relevance to the task in hand. Instructions which are highly relevant to the task – in the present case, the compulsory steps in the procedure – will tend to avoid the use of indirect directives since linguistic indirectness could detract from or even obscure the primary goal. At other places in the protocol, however, such as the Lead, the Tail, and, if present, supplementary information provided in the Procedure, which are less crucial to the task, different verb forms can occur: *should* for recommendations or for expected outcomes, *can* and *may* for possibilities and choices, or expressions of desirability for optional but recommended steps (*it is a good idea to…).*

A salient discursive feature of the standard protocol is its lack of any contextual anchoring, manifested by features such as: anonymity – the author's name is seldom mentioned, only the name of the laboratory or company – and impersonality, with practically no first person pronoun use (*I* or *we*), and absence of any interactional resources; atemporality – the protocol is presented as valid for any temporal framework; and lack of indexicality – no specific circumstances are mentioned. Standard protocols can therefore be considered 'idealized' texts, which give timeless, neutral, and typified instructions for action.

3. Corpus

The Protocols section of the OWW website currently contains 370 protocols, divided into four groups: protocols in development (9), In Vivo (104), In Silico (52), and In Vitro (205). For the present study, 40 experimental protocols were selected, 20 each from the In Vivo and In Vitro sections; In Silico protocols were not included, as they contain software rather than experimental procedures. A random selection was made, excluding those that were stubs, redirects, dewikified texts, or that closely duplicated a protocol that had already been selected in the other section. A smaller corpus of 20 print protocols (.pdf or Word documents) on highly similar topics was compiled from the sites of British and American biological engineering laboratories. The two subsets will be referred to respectively as OWW and Print.

For all 60 protocols, the text of the protocol page was collected and analysed using the concordancer AntConc 3.2.1; in addition, for the OWW subset, the following information was also collected: the revision history (number of edits, number of different contributors, comments on the edits), the talk pages, and the user profiles of the contributors. Table 1 gives an overview of the data.

	OWW	Print
Number of protocols	40	20
Total number of word tokens	25,916	8,659
Total number of word types	3,429	1,907
Average numbers per protocol:		
Edits	22.5	N/A
Contributors	4.8	N/A
visuals	0.5	0

Table 1. Overview of the corpus.

4. Comparative analysis

This section compares the two subsets, OWW and Print, focussing on structure (4.1), communicative purpose (4.2), and linguistic form (4.3).

4.1. Structure

The internal structure of the protocols in the two subsets was first compared, using the schematic structure given in Figure 2 above. As Figure 3 shows, OWW protocols have the same schematic structure of *Lead* > *Materials* > Procedure > *Tail*, presented in the same order, as print protocols. Migration to the web does not appear, therefore, to have radically modified what one could call the skeleton of the genre. There is however an appreciable expansion of the initial and concluding optional sections (*Lead* and *Tail*) in the OWW protocols: while only 40% (n = 8/20) of print protocols contain these sections, they feature in 87.5% (n = 35/40) of the OWW protocols.

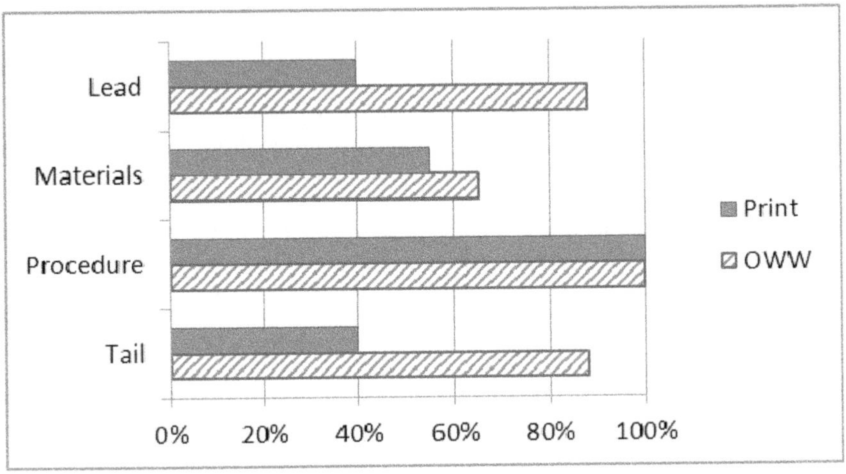

Figure 3. Comparison OWW/Print protocols – Schematic Structure.

In the print protocols, the brief *Lead* section mainly gives the rationale or use of the procedure (five occurrences), with one occurrence each of precautions to take, a description of the text structure of the protocol itself, and a reference to the assay kit used. In the OWW protocols, this section is not only longer on average, but has a wider variety of functions, as the following list makes clear: rationale or use; background history of the protocol and its motivation (such as problems encountered with the standard protocol); advantages of the protocol (fast, easy, cheap, etc.); scientific principle of the method; summary or overview of the procedure; hyperlinks to related or more specific protocols on the OWW site or elsewhere; links to outside sites or other OWW Discussion pages on the topic; recommendations; literature reference to the standard protocol of which it is a variant; priority claim; personal experience of the contributor; and origin of the name of the protocol. As this list indicates, the *Lead* in OWW protocols provides a much richer contextualisation of the protocol than in the typified Print versions. Examples (1) to (3) illustrate this feature:

(1) This protocol produces very clean, concentrated DNA, usually on the order of several mg/mL. It is labor intensive. Expect to spend most of the day doing it, but I like to store my plasmids in this form. (OWW-6)

(2) Native lysis is a staple protocol in practically every biochemistry lab, yet there is significant variability in the procols [sic] I have read. […] Besides, the machine is scary, lysis can be variable, and cleanup is a hassle. […] Be careful, if your protein is ~14 kDa you may inadvertently purify the added lysozyme instead of your target protein. Don't laugh, it's happened more than once and I have met people who solved the crystal structure of lysozyme by mistake. (OWW-14)

(3) I was in graduate school and had to make several codon substitutions at the same position in a cloned gene. I came up with this protocol and called it 'Around-the-Horn', a phrase used in baseball meaning to throw the ball through all of the base positions in a circle. The phrase originally referred to 'Rounding Cape Horn of South America' a very difficult, but rewarding journey ('Round-the-horn' is not difficult). Subsequently, a group published a paper describing the process, but I was first, take full credit, and have used it in ways not described by others. (OWW-24)

Similarly, the *Tail* section is absent or brief in Print protocols, which contain only four cases of literature references and five sets of comments or tips, whereas the OWW protocols feature lengthier and more varied types of concluding material: a large number of links to literature references; signed comments, tips, description of failed and successful experiments by various contributors (their OWW User pages accessible via dynamic links); explanation of the background thinking behind the protocol; links to related OWW pages and specific protocols; external links to discussions on the topic; links to the open source Biocoder version of the protocol; visuals (graphs of data obtained using the assay); estimation of the cost; trouble-shooting; and variants on the protocol. Examples (4) to (7) illustrate some of these features:

(4) Although not necessary for a protocol to work it is often useful to know the thinking behind the method development, this may help a user tailor the method […] or aid trouble shooting. After watching too much CSI I saw that gDNA was surprisingly stable at room temperature […]. (OWW-1)

(5) When is the reaction done? I never found a good answer for this in the literature. […] I routinely make three separate measurements. (Reshma; OWW-3)

(6) Frequently, people use disposable plastic cuvettes to measure both the culture turbidity and the yellow o-nitropenol. It is my experience that the disposable cuvettes have HORRIBLE optics: yes, they are clear, but the light path is pathetically distorted […]. The values can vary GREATLY even for the same

culture measured in different cuvettes. I recommend using […]. I have found my error in pipetting 150 µL of culture for Abs_{600} measurements was WAY less than using disposable cuvettes. (OWW- 3)

(7) You may want to read this discussion of what LiCl and sodium acetate are doing in this protocol. (OWW-6)

When the whole structure of an OWW page (see the partial screenshot in Figure 4) is compared to that of a print protocol (Figure 1), it can be seen that, in addition to these expansions of the schematic structure of the protocol itself, the online protocols are embedded in a completely different textual environment. The format is that of a typical wiki page, comprising: a sidebar of links (general navigation links, three research links, a toolbox, and links to the User pages of the Contributing Authors), and a search box; three tabs giving access to Talk, View source, and History pages; a Contents box of dynamic links to the different sections of the text; links in the text to information outside the page and/or the OWW site; hierarchical links in the Categories box; and information about the page (date last modified, number of hits) and the site. Two additional buttons enable newcomers to join OWW, and registered users to log in in order to make edits. On this single page, 62 hypertextual links can be activated. This format and the functions of its component parts are familiar to the reader – not because they form part of the protocol genre, but because they are part of the digital environment. The wiki text structure is trans-generic, in that it can be applied to many different genres – encyclopedias, collaborative fiction, collaborative journalism, online catalogues, cookery recipes, city websites – and endows any document, whatever the genre, with the same structural and interactive features; it is a technological affordance provided by the software. Just as online scholarly journals incorporate new Internet-enabled features and functions which bring added value to the traditional RA genre (Luzon 2007), the wiki format and, more generally, the digital medium endow OWW protocols with a host of additional features. Askehave and Nielsen argue convincingly that due to the importance of the interplay between genre and medium in web-mediated texts, the characteristics of the medium should be considered to be inherently part of the online genre: "media properties influence both the purpose and form of web-mediated genres and should therefore be included in the genre identification" (2005: 128). In the case of the OWW protocols, while the structural skeleton of the parent genre is still clearly visible and functional – the instructional core of the genre (Materials and Procedure) has migrated successfully to the web – one cannot account for the differences between the print and online versions if the characteristics of the medium are ignored. Hypertextuality undermines fixed generic structure, since the ways in which the protocol page can be read and used are numerous and highly variable.

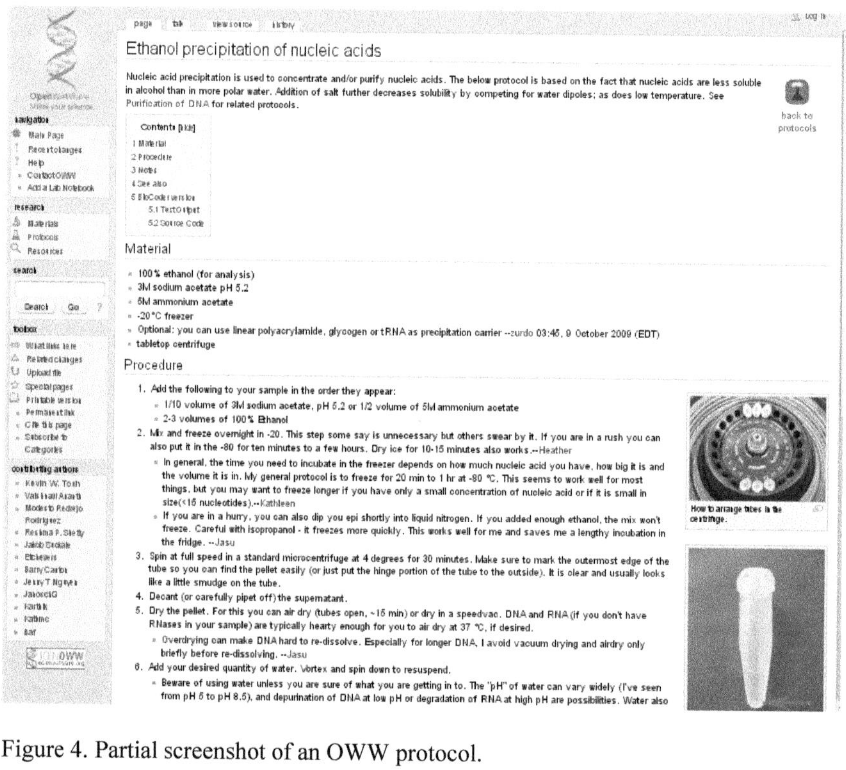

Ethanol precipitation of nucleic acids

Nucleic acid precipitation is used to concentrate and/or purify nucleic acids. The below protocol is based on the fact that nucleic acids are less soluble in alcohol than in more polar water. Addition of salt further decreases solubility by competing for water dipoles; as does low temperature. See Purification of DNA for related protocols.

back to protocols

Contents [hide]
1 Material
2 Procedure
3 Notes
4 See also
5 BioCoder version
5.1 Text Output
5.2 Source Code

Material

- 100% ethanol (for analysis)
- 3M sodium acetate pH 5.2
- 5M ammonium acetate
- -20°C freezer
- Optional: you can use linear polyacrylamide, glycogen or tRNA as precipitation carrier --zurdo 03:45, 9 October 2009 (EDT)
- tabletop centrifuge

Procedure

1. Add the following to your sample in the order they appear:
   - 1/10 volume of 3M sodium acetate, pH 5.2 or 1/2 volume of 5M ammonium acetate
   - 2-3 volumes of 100% Ethanol
2. Mix and freeze overnight in -20. This step some say is unnecessary but others swear by it. If you are in a rush you can also put it in the -80 for ten minutes to a few hours. Dry ice for 10-15 minutes also works.--Heather
   - In general, the time you need to incubate in the freezer depends on how much nucleic acid you have, how big it is and the volume it is in. My general protocol is to freeze for 20 min to 1 hr at -80 °C. This seems to work well for most things, but you may want to freeze longer if you have only a small concentration of nucleic acid or if it is small in size(<15 nucleotides).--Kathleen
   - If you are in a hurry, you can also dip you epi shortly into liquid nitrogen. If you added enough ethanol, the mix won't freeze. Careful with isopropanol - it freezes more quickly. This works well for me and saves me a lengthy incubation in the fridge. --Jasu
3. Spin at full speed in a standard microcentrifuge at 4 degrees for 30 minutes. Make sure to mark the outermost edge of the tube so you can find the pellet easily (or just put the hinge portion of the tube to the outside). It is clear and usually looks like a little smudge on the tube.
4. Decant (or carefully pipet off) the supernatant.
5. Dry the pellet. For this you can air dry (tubes open, ~15 min) or dry in a speedvac. DNA and RNA (if you don't have RNases in your sample) are typically hearty enough for you to air dry at 37 °C, if desired.
   - Overdrying can make DNA hard to re-dissolve. Especially for longer DNA, I avoid vacuum drying and airdry only briefly before re-dissolving. --Jasu
6. Add your desired quantity of water. Vortex and spin down to resuspend.
   - Beware of using water unless you are sure of what you are getting in to. The "pH" of water can vary widely (I've seen from pH 5 to pH 8.5), and depurination of DNA at low pH or degradation of RNA at high pH are possibilities. Water also

How to arrange tubes in the centrifuge.

Figure 4. Partial screenshot of an OWW protocol.

The web-mediated genre also differs in communicative purpose, as the following section argues.

4.2. Purpose

The concept of genres as 'social action' (Miller 1984) has stressed that a genre corresponds to a set of shared communicative purposes, and that discourse communities generate genres to answer their discursive needs and practices (Swales 1990). Reservations have been expressed, however (Askehave 1999; Askehave/Swales 2001), about the dangers of an exclusive focus on communicative purpose, and studies of web-mediated genres in particular argue that, given the multifunctionality of genres in this medium, a "downgrading of purpose as the guiding criterion for the determination of whether a text qualifies for membership in a given genre" is required (Garzone 2007: 26). This section will examine the notion of purpose more closely in the web-mediated protocols, again comparing them with the Print versions.

Print protocols are 'single-minded', in that their sole purpose is informative, to give procedural instructions on how to carry out the protocol. This instructional purpose is also a feature of the OWW protocols (see Figure 4), and in both cases stems from the fact that the community producing the protocols is composed of experimental researchers in biology who need and use protocols in their daily activity. The OWW community, however, is, in addition, an Open Science community, with an agenda of promoting transparency in science, data-sharing, discussion, and the collaborative production of texts. The choice of a wiki format corresponds to this agenda. Furthermore, the majority of the initial and subsequent contributors to OWW are junior researchers – PhD students or post-docs – who form a somewhat occluded sub-community within the research hierarchy. In studies of academic writing, it is generally considered that the dominant mode of learning to write for junior or novice researchers is by legitimate peripheral participation (Lave/Wenger 1991), during which they gain proficiency in the research genres of their community through practice, imitation, and guidance from senior researchers (Belcher 1994). In the case of the Web-mediated protocols, however, these junior researchers have not only appropriated the existing genre, but have substantially modified it, introducing several innovations to make it fit their purposes and perceived needs: as many of the examples from the data show, the traditional print protocols are felt to provide unsatisfactory, insufficient or sometimes downright wrong information. The web provides a suitable channel for this re-purposing of the protocol genre by reason of the greater 'expressivity' or outspokenness that is a recognised feature of the medium. One should bear in mind, however, that while protocols are an essential tool in laboratory practice, as texts they are a non-prestigious scientific genre; the stakes are therefore much lower than if the junior researchers had undertaken to re-purpose the research article, for instance.

Nevertheless, it seems clear that the OWW protocols – unlike print protocols – are not single-purpose but multi-purpose, and moreover that the combination of an Open Science agenda and the affordances of the web medium have resulted in a range of novel features that seriously question the stability of the traditional genre. The common point of these innovations concerns the indexicality of research, i.e. the contextual contingencies of scientific activity. As long ago as 1978, Knorr and Knorr pointed out that the real work that goes on in the lab involves tinkering, indeterminacy and *ad hoc* solutions to unexpected problems, all of which are ironed out in the typified abstraction of the published research article, where the 'practical' frame of lab work is transformed into the 'scientific' frame of the RA. Practical know-how, or lab lore, along with failed – and therefore unpublished – experiments remained 'dark data'. Print protocols, though highly practical, reveal very little of this empirical lab lore in fact: rather like another instructional genre, cookery recipes, they give a typified or idealized version of the procedure, omitting much of the tacit knowledge and hands-on experience required to make it work. Much of this insider information is revealed, however, in the web-mediated genre: the Open Science movement has not only opened up collaboration and data-sharing between scientists, it has also opened a window on scientific practices. Examples (8 to 18) give a brief illustration of this.

Standard protocols do not work or do not give the expected results: for this reason, many of the OWW protocols are adaptations and improvements on the standard ones. OWW-3, for example, is entitled 'A better Miller' (the Miller protocol being the standard one).

(8) Some protocols call for placing the cells on ice during lysis [...] This never made much sense to me (OWW-14)

Failed or unsuccessful attempts are honestly recounted:

(9) The direct addition of crushed material [...] wasn't giving repeatable clean products [...] I could get a green juice out but not much luck with PCR products (OWW-1)

(10) I have had reactions that just wouldn't produce product (for whatever reason). (OWW-24)

Many lab procedures are real drudgery:

(11) attempting to purify the band by cloning all the reaction products [...] can be extraordinarily tedious. (OWW-33)

Lab lore is empirical knowledge:

(12) This has to be determined empirically, as does the amount of antibody required. (OWW-4)

(13) It should be pretty cloudy but not super dense. I just do it by eye. (OWW-10)

Tinkering: the equipment used is often what comes to hand. In addition to information in the protocol itself, the User pages of contributors contain further tips, called 'Bench Stuff', 'Cruft', 'Cautionary Tales', etc.

(14) all of the materials came from regular shops, including Radioshack and an Asian grocery store in Sacramento, CA. (OWW-23)

Scientists are human!

(15) I use 250-500uL, depending on how much I have and how lazy I am.

(16) Many times, I have forgotten to phosphorylate the primers before doing the PCR. (OWW-24)

Differences of opinion are also openly thrashed out, often by including in the protocol itself the diverging opinions of different contributors:

(17) The original paper *chung* suggests freezing the cells immediately using a dry ice bath. I (BC) have used liquid nitrogen quite successfully instead of dry ice. Simply placing the cells at – 80°C also seems to work well. (Jkm)

(18) You might consider using a syringe filter rather than spins to do the washes when preparing the cells. This can apparently save a lot of time, though I've yet to try it myself. (Jason R. Kelly) I tried with a 0.22μm syringe filter and 50ml OD 600 e.coli. The filter kept jamming, not very successful. Perhaps a larger, vacuum filter would be better? (hugh kingston). (OWW-5)

As (17) and (18) show, the collaborative co-construction of the protocols is very much in evidence; even decisions to keep or remove a protocol on the site are made collectively, as shown by several exchanges on OWW. As Myers comments, "a sense of community is absolutely essential" to wikis (2010: 25).

The indexical features of scientific activity revealed by examples (8) to (18) bear a remarkable resemblance to the glimpses of lab life encountered in a spoken scientific genre, namely conference presentations (Rowley-Jolivet 2002). Nor does the resemblance stop there: from a linguistic point of view, OWW protocols, though written, manifest many features typical of oral discourse, as shown in the following section.

4.3. Linguistic features

As Giltrow and Stein remark, due to the low degree of conventionalization of form-function relationships generally observed in Internet genres, "studies of the linguistic markers of Internet genres are few and far between" (Giltrow/Stein 2009: 11). By comparing the linguistic features of the Print and OWW subsets analysed here, however, it seems possible to propose a characterization of the web-mediated texts, though for reasons of space only a few aspects can be examined.

	Print	OWW
Verb forms		
Imperatives	73.5%	36.5%
Present tense	17.5%	38.6%
Modals	8.5%	12.5%
Other tenses	0.5%	12.4%
Total	100%	100%
Personal pronouns		
I, me, my	4	221
you, your	27	357
we, our	1	29
Total	32	607

Table 2. Verb forms and personal pronoun subjects in Print and OWW protocols.

As can be seen in Table 2, standard protocols use a highly restricted range of verb forms: imperatives are by far the most dominant form, accounting for almost three-quarters of the total; practically the only tensed form used is the present (the 'other tenses' category comprises a mere six occurrences of *have –EN*, all in temporal clauses with *when/after*); a small proportion of modals are used for recommendations, expected results, etc., as in example (19):

(19) Place remaining agarose in 65°C oven and allow gel to cool for 10 to 15 minutes. The liquid in the gel form will shrink and gel should be 'topped off' with more agarose solution. (Print-7)

In the OWW subset, the imperative is still used for the main instructional steps of the procedure; the texts are therefore easily recognizable, on purely internal evidence, as belonging to the protocol genre. These instructions are, however, interspersed with a large number of comments and framed, as discussed in section 4.1 above, by greatly expanded *Lead* and *Tail* sections. As a result, the imperative represents only a little over one-third of all verb forms, indicating that the texts are not solely instructional but have a wider range of functions; these are expressed via tensed forms, which account for over 50% of the total. Present tenses (38.6%) are regularly used to describe the contributor's usual practice (example 20), or to render the immediacy of the operations (example 21), while the presence of past tenses (8%

of the total) indicates that the web-mediated protocols include a substantial narrative element (example 22). Relating this to the recent shift in narrative analysis away from narrative as text to narrative as practice within social interaction (de Fina/Georgakopoulou 2008), it can be suggested that in the OWW protocols, narrative contributes to establishing a sense of online community among contributors who are geographically widely dispersed: users of the site will recognize themselves and their own problems in the episodes recounted in the texts, and the use of personal pronouns (see Table 2) further enhances this interpersonal dimension.

(20) I can make pretty close to a 10 mg/mL solution by just adding 400 μL of buffer to the tube. So, I no longer weigh out lysozyme each time, I just assume that my tip carries about 4 mg of protein.

(21) If you're adding a control RNA for quantification, now is when you do it.

(22) I started using this back when I was running into trouble with 3-way assembly and thought this might save a step if it worked.

As several of the examples cited in this chapter show, the lexico-grammatical register of many passages in the OWW protocols is often close to that of conversational exchanges in the lab or to the informal style of emails and blogs. This is manifested in the frequent use of features such as: capital letters to indicate prosodic stress (*HORRIBLE optics*); exclamation marks (*clone away! Smile because you didn't have to work with the loud-ass sonicator!*); contractions (*don't laugh, I've, it's*), ellipsis of the pronoun subject (*Recently tried this again…, can switch to Falcon tubes here*) and run-on comma structures; spoken discourse markers (*actually, now, well, so*); non-standard forms for humorous effect (*if they ain't dead yet, they ain't gonna be*); or colloquial lexis (*any other junk that may be in your sample*, *they absorb like crazy, pretty gooey*). In the Print protocols, while a small number of capitals are used for important safety warnings and occasional verb contractions are found in the comments, the register used is standard written English. In the OWW protocols, in contrast, along with the indexicality of lab work, the style of lab talk has also migrated to the web. A further typical feature of lab exchanges – their humour – is also very common on the User pages of OWW, but cannot unfortunately be presented here.

5. Concluding remarks

This study has highlighted some of the transformations that a well-established scientific genre, the laboratory protocol, has undergone in its migration from the traditional print to the web-mediated version. While the basic skeleton of the genre subsists in the online protocols, substantial changes have taken place in the structure, the purpose and the linguistic expression: the structure has been considerably expanded, the purpose diversified, and certain lexico-grammatical features radically altered, under the dual influences of the medium itself and the Open Science agenda. The web appears to happily accommodate a much greater degree of variation in generic features than print media, but a certain tension can be discerned between the initial characteristics of the genre and these novel features: will there come a point where these expansions and innovations mean that the web-mediated protocols cease to be considered as examplars of the protocol genre? It remains to be seen whether similar transformations will occur with more prestigious research genres as they adapt to the new communicative context and take advantage of its affordances.

References

Askehave, Inger 1999. Communicative purpose as genre determinant. *Hermes, Journal of Linguistics* 23, 13-23.

Askehave, Inger / Swales, John M. 2001. Genre identification and communicative purpose: a problem and a possible solution. *Applied Linguistics* 23/2, 195-212.

Askehave, Inger / Ellerup Nielsen, Anne 2005. Digital genres: a challenge to traditional genre theory. *Information Technology and People* 18/2, 120-141.

Belcher, Diane 1994. The apprenticeship approach to advanced academic literacy: graduate students and their mentors. *English for Specific Purposes.* 13, 23-34.

Borgman, Christine L. 2007. *Scholarship in the Digital Age: Information, Infrastructure and the Internet.* MIT Press.

Bradley, Jean-Claude 2006. <http://drexel-coas-elearning.blogspot.com/2006/09/open-notebook-science.html>.

Brown, Penelope / Levinson, Stephen C. 1978. *Politeness: Some Universals in Language Usage.* Cambridge: Cambridge University Press.

Crowston, Kevin / Williams, Marie 1996. Reproduced and Emergent Genres of Communication on the World-Wide Web. <http://florin.syr.edu/~crowston/papers/webgenres.html,>.

David, Paul A. 2004. *Towards a Cyberinfrastructure for Enhanced Scientific Collaboration: Providing its 'Soft' Foundations May Be the Hardest Part.* Research report N° 4, Oxford Internet Institute.

de Fina, Anna / Georgakopoulou, Alexandra 2008. Narrative analysis in the shift from texts to practices, Introduction to Special Issue of *Text & Talk.* 28/3, 275-281.

Garzone, Giuliana 2007. Genres, multimodality and the World Wide Web: theoretical issues. In Garzone, Giuliana / Poncini, Gina / Catenaccio, Paola (eds) *Multimodality in Corporate Communication: Web Genres and Discursive Identity.* Milano: Franco-Angeli, 15-30.

Garzone, Giuliana / Poncini, Gina / Catenaccio, Paola (eds) 2007. *Multimodality in Corporate Communication: Web Genres and Discursive Identity.* Milano: FrancoAngeli.

Giltrow, Janet / Stein, Dieter (eds) 2009. *Genres in the Internet. Issues in the Theory of Genre.* Amsterdam: John Benjamins.

Jankowski, Nicholas W. 2007. Exploring e-science: An introduction. *Journal of Computer-Mediated Communication.* 12/2, article 10.

Kepler, Thomas B. / Marti-Renom, Marc A. / Maurer, Stephen M. / Rai, Arti K. / Taylor, Ginger / Todd, Matthew H. 2006. Open Source Research – The Power of Us. *Australian Journal of Chemistry* 59/5, 291-294.

Knorr, Karin D. / Knorr, Dietrich W. 1978. *From Scenes to scripts: on the Relationship between Laboratory Research and Published Paper in Science*. Vienna: Institut für Höhere Studien.

Lave, Jean / Wenger, Etienne 1991. *Situated learning: legitimate peripheral participation*. Cambridge: Cambridge University Press.

Luzón, Maria J. 2007. The added value features of online scholarly journals. *Journal of Technical Writing and Communication.* 37/1, 59-73.

Merton, Robert K. 1973. *The Sociology of Science: Theoretical and Empirical Investigations*. Chicago: University of Chicago Press.

Miller, Carolyn 1984. Genre as social action. *Quarterly Journal of Speech.* 70, 151-167.

Murcia-Bielsa, Susana 2000. The Choice of directive expressions in English and Spanish instructions: A semantic network. In Ventola, Eija (ed.) *Discourse and Community: Doing Functional Linguistics*. Tübingen: Gunter Narr, 117-146.

Myers, Greg 2010. *The Discourse of Blogs and Wikis*. London: Continuum.

Nentwich, Michael 2003. *Cyberscience: Research in the Age of the Internet.* Vienna: Austrian Academy of Sciences.

Orlikowski, Wanda J. / Yates, JoAnne 1994. Genre repertoire: The structuring of communicative practices in organizations. *Administrative Sciences Quarterly.* 33, 541-574.

Peikola, Matti / Skaffari, Janne / Tanskanen, Sanna-Kaisa (eds) 2009. *Instructional Writing in English: Studies in Honour of Risto Hiltunen.* Amsterdam: John Benjamins.

Rowley-Jolivet, Elizabeth 2002. Science in the making: Scientific conference presentations and the construction of facts. In Ventola, Eija / Shalom, Celia / Thompson, Susan (eds) *The Language of Conferencing*. Frankfurt: Peter Lang, 95-125.

Swales, John M. 1990. *Genre Analysis*. Cambridge: Cambridge University Press.

Waldrop, M. Mitchell 2008. Science 2.0: Great New Tool, or Great Risk? Wikis, blogs and other collaborative web technologies could usher in a new era of science. Or not. *Scientific American.* January 9, 2008.

MARISTELLA GATTO

Centrifugal and Centripetal Forces in Web 2.0 Genres. The Case of Wikipedia

1. Introduction

The aim of this chapter is to explore the impact that the recent technologies we associate with Web 2.0 may have had on some specific genres. In particular, the chapter investigates the contribution of the technological affordances of Web 2.0 to the action of the centripetal and centrifugal forces of discourse (Bakhtin 1982), with reference to the case of Wikipedia, the well-known cooperative online encyclopaedia. Following this introduction, Section 2 briefly surveys recent research in genre studies with a special focus on issues relating to stability and change in web genres. Then, some key questions relating to Web 2.0 as a new writing space are discussed in Section 3, with a focus on such fundamental notions as *authorship vs audience, consumers/ producers vs prosumers, upload vs download*, *taxonomies vs folksonomies*, all considered critical to Wikipedia as a cooperative writing enterprise. Finally, in the Section 4, the evolution of genre in the specific case of the encyclopaedia is investigated by comparing/contrasting different stages in the history of a sample of Wikipedia entries, so as to provide evidence of the action of the centrifugal/centripetal forces at work in this prototypical Web 2.0 genre. The research hypothesis underlying the case studies reported in the final part of the chapter is that entries may not necessarily comply with generic conventions at the time of their creation, subject as they are to the centrifugal forces of individual creativity, but eventually do so in the long run by effect of the centripetal forces unleashed by the action of the Wikipedia community.

2. Between stability and change: contemporary approaches to genre studies

No attempt at approaching the issue of Web 2.0 genres can be made without recalling some theoretical concepts concerning genre theory in general and web genres in particular. By no means a new question, with attempts at a classification of communicative acts into genres dating back at least to Aristotle, genre has indeed always been centre stage in the discourse about human communication, and it comes therefore as no surprise that the advent of new media in the context of Computer Mediated Communication has proved "peculiarly inviting to discussion of genres" (Giltrow/Stein 2010: 1). As a matter of fact, the web itself enables a new communicative setting which reconfigures the conditions in which genres are produced, consumed and negotiated.

Particularly influential in the contemporary debate on digital genres have been Bakhtin's ideas on the centrifugal and centripetal forces of discourse. Shifting the discussion of genres away from its formalist basis at the beginning of the 20th century towards a more dynamic view of discourse, and enlarging the notion of genre so as to include non-literary genres and speech genres, Bakhtin put forward an idea of genre strongly affected by dialectic tensions. For him, all utterances have "definite and relatively stable typical forms of constructions", and from these primary (speech) genres the more complex secondary genres of writing are derived. Nonetheless, he acknowledged that despite involving regularities and typifications, generic forms are constantly sensitive to negotiation and struggle, as well as to the tensions of those that he labelled as the centripetal and centrifugal forces of discourse. "Every concrete utterance of a speaking subject", Bakhtin argues, "serves as a point where centrifugal as well as centripetal forces are brought to bear" (Bakhtin 1982: 272). The centripetal forces are forces that centralize discourse, forcing all the elements of language and all of its various rhetorical modes into one single form; the centripetal force of *monologia* thus results in a system of norms, in a sort of 'official' and unitary standard language. Centrifugal forces, on the other hand, combine with *heteroglossia*, and decentralize discourse, moving it towards multiplicity (Bakhtin 1982: 272ff). As a result of this dynamic view of discourse, genre is perceived as a relatively fragile category, much more "changeable, flexible and plastic" (Bakhtin 1986: 80) than other language patterns, and ultimately subject to constant change.

It was precisely drawing on Bakhtin's insight into discourse as a field of centripetal and centrifugal forces, and combining such a view with social theory that explores agency and structure as well as with linguistic studies of variation, that recent genre theory – especially in the context of the North American New Rhetorical School – has definitely foregrounded the dynamism, flexibility and change inherent to genres (Miller/Shepherd 2010: 264). Following Bakhtin, Berkenkotter and Huckin have thus argued that genres are "inherently dynamic rhetorical structures" and that they are "sites of contention between stability and change [...] constantly changing over time in response to the socio-cognitive needs of individual users" (Berkenkotter/Huckin 1995: 6).

Such a dynamic view of genres and the very notion of 'genre change', which is so prominent in contemporary genre theory, are also key issues in the debate concerning web genres. Nevertheless, as Miller and Shepherd (2010: 264) convincingly point out, "there's something problematic about the very idea of genre change", especially when the inherent flexibility and fluidity of genres meets technological affordances that enhance such instability and changeability, as is the case with new digital media. Genre change, they suggest,

> problematizes precisely what makes genres generic. [...]If a genre is a mark of recurrence, what is it that recurs, especially in a setting of dazzling volatility as the internet? How do communities create collective typifications when neither technology nor culture will hold still? What is it that can be reproduced sufficiently to create genre identity in varying instantiations? Can there be a stable or identifiable dimension of a genre as it is adapted from one medium to another?

As a matter of fact, web texts and web genres are not simply more complex and unpredictable than their paper counterparts, but have actually been recognized as being more prone to the effect of centrifugal forces which are the cause of constant genre evolution (Yates/Sumner 1997: 3;

Santini 2005: 1). As a result, the web scenario in terms of genre is extremely varied. On the one hand, it is clear that the web hosts a number of traditional genres which have simply changed medium – from paper to electronic – without any further modification of their intrinsic features (reproduced genres); on the other hand there are text-types and genres that are undergoing more complex processes of 'remediation' which make them definitely hybrid genres (adapted novel genres); finally there are completely new – often highly individualized – genres (emerging genres) (Santini 2007). Typical examples of the above can be found in the abundance of literary works that have been made available through the web, which have only changed medium (from paper to digital) without changing any of their properties (reproduced genres), or the Internet version of newspapers, which are being adapted to the new medium to take advantage of new functionalities (adapted novel genres), or emerging genres like the homepage or the blog.

This varied scenario has ultimately foregrounded in crucial ways the relationship between genre and medium. This means that it has been increasingly acknowledged that the medium adds unique properties to web genres in terms of production, function and reception, which cannot be ignored in genre characterization (Askehave/Ellerup Nielsen 2005). It has also been argued that all attempts at characterizing web genres have been so far seriously handicapped by a relatively simple notion of genre that has only minimally been extended from traditional non-multimodal conceptions (Bateman 2008: 197). New models for genre categorization are therefore being explored, especially in the context of computational linguistics, to account for the different forms of interaction allowed by the medium in digital genres, and research in the field reveals a debate that is still "thrilling and heated" (Santini/Mehler/Sharoff 2010: 8).

As to the apparent stability of traditional genres, compared to a more evident constant evolution of digital genres, it might appear quite obvious how the perceived stability of traditional print publications is to some extent also related to the medium itself, and reinforced by mechanisms of production, distribution and circulation of printed texts, which are connected to hegemonic powers within society (e.g. publishing houses; editorial boards; disciplinary communities etc.). By contrast, the recent development of web technologies that enable immediate publication and online distribution of user-generated content, thus blurring any distinction between production and consumption, has certainly had a destabilizing effect on genres. Nonetheless, the democratization of genres enabled by digital technology has by no means reduced the supporting function of genres for particular communicative needs and work practices. As Yates and Sumner aptly argued:

> Current technology *has* destabilised existing genres compared to print technology. However, the result is not a breakdown or loss of recognisable genres. Instead, the distinction between producers and consumers of digital documents is being blurred and we are seeing the democratisation of genre production. More people are providing more input (either implicitly or increasingly explicitly) into the creation and production of genres. As a result, over time communities evolve increasingly well-defined genres to better support their particular communicative needs and work practices. In effect, rather than seeing a lessening in the role of genre, *the new burden for providing fixity in communications is being met by increased reliance on genre.* (Yates/Sumner 1997: 3).

It is precisely this supposed effect of democratization and challenge of genres resulting from the centrifugal forces unleashed by the new media, and the simultaneous reliance on genre, as the 'new burden of fixity' due to the action of centripetal forces which are nonetheless at work in these new media, that the present chapter aims to explore by instance of Wikipedia.

3. A prototypical Web 2.0 genre: the case of Wikipedia

Investigating in detail the effect of the technological affordances that make up the so-called Web 2.0 on the encyclopedia genre requires an intermediate step to focus on what it actually means to communicate in a Web 2.0 environment. As a famous video by anthropologist Michael Wesch illustrates, Web 2.0 technologies have basically made the creation and distribution of user-generated content easy and free for everybody, thanks to a change in the code (the example Wesch gives is HTML replaced by XML) that enables the separation of form and content. In so doing, this technology has however also raised fundamental issues which are relevant to communication and society at large, ultimately contributing to new forms of interaction between man and the machine and among human beings themselves (Wesch 2007).

Implicit in this representation of the meaning and implications of Web 2.0 is the obvious realization that social computing practices have had a deep impact on the production, modification and dissemination of texts and hence information and knowledge. Particularly significant is the role played by wikis as prototypical Web 2.0 tools, of which Wikipedia is the most significant application. By allowing real-time publication of individual content without any previous editorial revision, wikis have unleashed a number of centrifugal forces which have had as a consequence, among others, an increasingly dynamic merging of the otherwise apparently distinct functions of audience and authorship. Furthermore, the ability for users to edit and re-edit web pages in real time actually displays the inescapable instability of texts, while de-emphasizing the role of individual authorship. Wikis thus provide material evidence of the dynamic and inherently social nature of language, and allow discourse to emerge that is continually negotiated and articulated through a community of users (Ray/Graeff 2008: 39-40).

Such a complex relationship between author and audience is actually foregrounded in any Wikipedia article through what could be termed a hypermediacy effect (Askehave/Ellerup Nielsen 2005). In any Wikipedia page, the graphic interface itself lays bare the continual interrelation between 'author', 'reader', and 'editor' functions and these otherwise separate roles are materially interconnected in one single entity, that is the threefold set of Article, Discussion and History which together make up a Wikipedia entry:

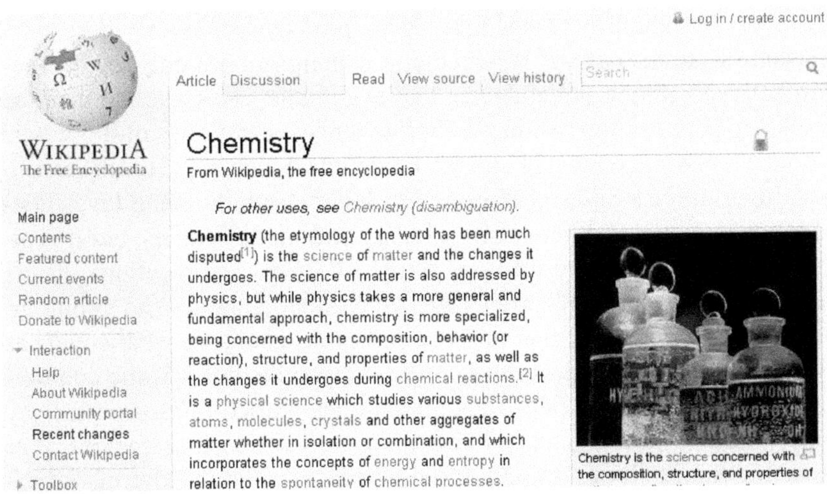

Figure 1. Wikipedia entry for 'Chemistry'.

Indeed the coexistence on the very graphic interface of such different and yet complementary forms of textuality as 'article' and 'discussion', along with the correlate functions 'read' and 'edit', as well as the possibility to re-view the 'history' of each single article, clearly foregrounds patterns of negotiation and conflict in the production of knowledge content. They also significantly lay bare the actual process behind a cooperative writing enterprise, and behind the collective creation of knowledge, showing the limits and fallacy of any unproblematized notion of individual authorship (Ray/Graeff 2008). As a consequence, the monologic appearance of the encyclopedia entry as a genre, exemplified at its best in its printed form, is turned by Wikipedia into a fluid multiple perspective dialogic object thanks to the technological affordances of the new medium.

This blurring of clear-cut distinctions between authors and readers in Wikipedia is also related to changes which have invested the realm of the economy and have been investigated with reference to society at large. One such change is the apparent end of any separation between consumers and producers, with the emergence of the new category of *prosumers* (Toffler 1981), a blend of the words *producer* and *consumer* suggesting a new economic relationship in which no distinction is made between those who produce value and those who consume it. In the case of Wikipedia, it is therefore not only the distinction between author and audience that eventually fades out, but also the boundary between the two more comprehensive categories of those who produce knowledge content and the potential consumers. This is certainly going to have long-term effects, which are currently being investigated at the convergence between sociology and media studies, but it is also having an immediate effect in terms of generic conventions. In the decentralized process of 'produsage' of Wikipedia there is no strong influential editorial board that makes *a priori* decisions on what counts as knowledge worth being stored in an encyclopedia, but it is the community as a whole that creates new taxonomies (called *folksonomies*)[1] by simply adding new content or providing tags to existing content, thus replacing the traditional expert-based paradigm of knowledge classification according to fixed schemata (Bruns 2008: 191).

This inevitably leads to considerations concerning the imbalance between upload and download as a recent phenomenon on the web. Indeed, only a few years ago, as Kevin Kelly argued in a very influential article on *Wired* (2005), download rates definitely exceeded upload rates, the basic assumption being that ordinary people had no need to upload. Ordinary people were still considered as mere consumers of web content, not producers. Today, on the contrary, all sort of users upload content on the web, an interaction that assumes participation, or 'prosumption', rather than mere consumption. With reference to Wikipedia this means that the encyclopedia is no longer only the place where people (as readers and consumers) go to absorb other people's knowledge, but also the place to go (as authors and producers) to contribute their own knowledge.

As to the impact that all this has on the specific issues of genres, it goes without saying that the 'generic integrity' of Wikipedia as an encyclopedia is deeply affected by the fact that its contributors' writing ability and levels of knowledge may vary greatly, as does their competence with reference to genre conventions. Not only is there the well-known problem of potentially undesirable authors who can con-

1 The word 'folksonomy' is a neologism formed by 'folks' and 'taxonomy' coined in 2004 by Thomas Vander Wal (see Peters 2009: 154).

tribute as easily as 'good' authors, which often results in acts of vandalism at the level of knowledge content, but the very formal consistency of Wikipedia is seriously impaired by the apparently uncontrolled input of multiple users. Not all users can be considered as 'experienced users' of the encyclopedia genre, and yet they are nonetheless allowed to produce and publish within this specific genre – often unconsciously breaking its established conventions.

As a community, however, Wikipedia has addressed these problems through a number of key features which are embedded in, and allowed by, technology itself, and which have an effect at the level of the preservation of the generic integrity of Wikipedia as an encyclopedia. The most interesting features are the existence of Wikipedia Roles, of Discussion Pages, of a Style Manual, and of Featured Articles. Such features contribute to the generic integrity of Wikipedia in a way which is not dissimilar to the action described by Bhatia for the peer review process or editorial intervention (Bhatia 2004: 189-190).

The very existence of a restricted number of Wikipedia users who have administrative privileges and can restore or block specific entries when this is deemed necessary to counteract acts of vandalism, suggests that forms of power and control are extant even in the apparently anarchic realm of social computing practices. Secondly, the auxiliary wiki objects called "Discussion pages" which accompany each Wikipedia article, suggest that each communicative act in Wikipedia is subject to a continual peer review process. Finally, the style manual indicating a number of guidelines which the authors of Wikipedia entries should follow, along with the nomination of Featured Articles, that is articles which are declared by the Wikipedia community to be of a particularly high level and hence represent a model, all contribute to reducing diversity and potential anarchy.

Particularly interesting in this context are the criteria on the basis of which featured articles are nominated. Featured articles are recognized by the community as being:

> (a) *well-written:* its prose is engaging, even brilliant, and of a professional standard;
> (b) *comprehensive:* it neglects no major facts or details and places the subject in context;

> (c) *well-researched*: it is a thorough and representative survey of the relevant literature. Claims are verifiable against high-quality reliable sources and are supported by online citations where appropriate;
> (d) *neutral:* it presents views fairly and without bias; and
> (e) *stable:* it is not subject to ongoing edit wars and its content does not change significantly from day to day, except in response to the featured article process.[2]

Surprisingly, these criteria have been found as not dissimilar from those that could be mentioned for printed encyclopedias (see Crawford 2001; Stvilia *et al.* 2005). The only exception is found in 'stability', a dimension that would be useless to require from traditional printed encyclopedias, which have no inherent aptitude to change once they have been published, but which is obviously a parameter to consider for an online cooperative encyclopedia. It comes anyway as a surprise that being 'stable' should be valued as a quality criterion in the context of Wikipedia where – it could be assumed – it is precisely the opposite (the possibility of continual change) that should be considered as a guarantee for good quality. It is as if, despite the revolution brought about by the medium, Wikipedia articles still aspired to a condition of stability, permanence, fixity – while remaining inherently and constitutionally open to change.

Finally, for a fuller understanding of the special textuality of Wikipedia, one should mention the History flow as a peculiar property of Wikipedia texts. By saving cached files of previous content for each single entry, so that it can be reinstated in case someone erases or vandalizes its entire content, the Wikipedia community actually stores the whole process of creation of each single article, both in terms of its content and of its form.

Each entry in Wikipedia thus becomes a palimpsest (Ray/Graeff 2008: 40), virtually including all possible versions of that entry, and the stages of each entry can be explored in a sort of journey into the creation of knowledge content by multiple users which is at the same time a journey into the text that actually makes that content visible and accessible. In this palimpsest one can see to what extent the 'textual activity' is as important as the 'scientific activity', as

---

2 <http://en.wikipedia.org/wiki/Wikipedia:Featured_articles>.

Bhatia (1997: 363) would suggest, in the dissemination of knowledge. And it is this palimpsest that can be profitably explored to see the centrifugal/centripetal forces at work in Wikipedia, with specific reference to the problem of 'generic integrity'.

4. History flows: Centripetal and centrifugal forces in Wikipedia

In this final section the History flow of a sample of Wikipedia entries will be explored to provide textual evidence of genre-related issues in Wikipedia, as discussed so far. Examples will be taken from a miscellaneous collection of Wikipedia entries, as well as from a corpus of entries (WikiChem), for which a more comprehensive qualitative/quantitative analysis is still in progress.[3] The research hypothesis is that entries may not necessarily comply with generic conventions at the time of their creation, subject as they are to the centrifugal forces of individual creativity, and prone as they are to suffer from lack of generic competence on the part of users (as well as from apparent lack of 'gatekeepers'), but they eventually tend to do so in the long run by effect of the centripetal forces unleashed by the action of the Wikipedia community. Special attention will be given in Section 4.1. to the generic integrity of each article, focusing on the reaction it prompts in the community of Wiki-pedians in terms of adherence to their generic expectations, whereas in Section 4.2. the lead section of a sample of entries will be investigated in more detail with a focus on editorial changes at the lexico-grammatical and textual levels. No reference is made here to quality and accuracy in content, aspects extensively dealt with in such works as Stvilia *et al.* 2005 and Cusinato *et al.* 2009.

3 The WikiChem Corpus is 160,000 token corpus of Wikipedia articles about chemistry collected in September 2010. For each entry also the first version has been downloaded and the resulting corpus composed of the first versions of the entries collected for WikiChem makes up the 24,120 token WikiChem1 corpus.

4.1. Generic integrity and revisions

A telling example of issues concerning the debatable generic integrity of Wikipedia articles is found in the History of the article on 'Genre'. In the screenshot below, one can see the incipit of this entry at the time of its creation on 17/3/2002:

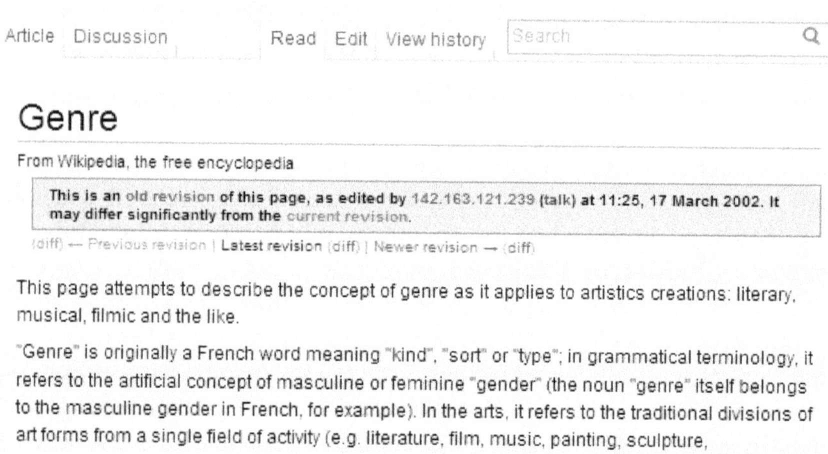

Figure 2. Wikipedia entry for 'Genre' (17/3/2002).

The opening sentence of this article ("This page *attempts to describe* the concept of genre as it applies to artistics [sic] creations: literary musical, filmic *and the like*") clearly does not comply at all with generic conventions which suggest that an encyclopedia entry should start with general information not dissimilar from a dictionary entry. On the contrary, defining the 'page' as an 'attempt to describe the concept of genre' suggests uncertainties on the author's part both as to the content of the entry and as to its form.

This is certainly the most evident effect of the empowerment, by means of technology itself, of users with no knowledge of genre conventions to publish unsupervised content – which prompts other users to intervene and preserve not only accuracy in content but also generic integrity. What follows is a screenshot from the History of 'Genre', showing in detail one stage of the revision process:

From Wikipedia, the free encyclopedia
(Difference between revisions)

Revision as of 12:27, 8 June 2004 (edit)	Revision as of 11:24, 31 October 2004 (edit)
Kenny sh (talk | contribs)	(undo)
m (→External links: +http://www.imdb.com/Sections/Genres/)	Violetriga (talk | contribs)
← Previous edit	(totally reorganised - needs expansion and article summaries)
	Next edit →

Line 1:	Line 1:
[[de:Genre]] [[es:género]] [[sv:genre]] [[da:genre]]	+ The term '''genre''' refers to the traditional divisions of [[fine art|art forms]] from a single field of activity into various kinds according to criteria particular to that form.
-	
-	
- This page attempts to describe the concept of '''genre''' as it applies to artistic creations: [[literature|literary]], [[music|musical]], [[cinema]]tic and so forth.	
"Genre" is originally a [[French language|French]] word meaning "kind", "sort" or "type"; in grammatical terminology, it refers to the artificial concept of masculine or feminine [[grammatical gender]] (the noun "genre" itself belongs to the masculine gender in French, for example).	"Genre" is originally a [[French language|French]] word meaning "kind", "sort" or "type"; in grammatical terminology, it refers to the artificial concept of masculine or feminine [[grammatical gender]] (the noun "genre" itself belongs to the masculine gender in French, for example).

Figure 3. Difference between revisions.

As the image clearly reveals, at one stage (revision as of 31/10/2004) in the History of this Wikipedia entry, one user deleted the clumsy opening sentence, and replaced it with a more conventional one in form, though no less elusive in content:

> The term genre refers to the traditional divisions of art forms from a single field of activity into various kinds according to criteria particular to that form. 'Genre' is originally a French word meaning 'kind', 'sort' or 'type'; in grammatical terminology, it refers to the artificial concept of masculine or feminine grammatical gender (the noun 'genre' itself belongs to the masculine gender in French, for example).

Indeed, if we compare these two early versions with a more recent entry (13/09/2010), we notice many changes that have been made to enhance its conformity with the style of an encyclopedia:

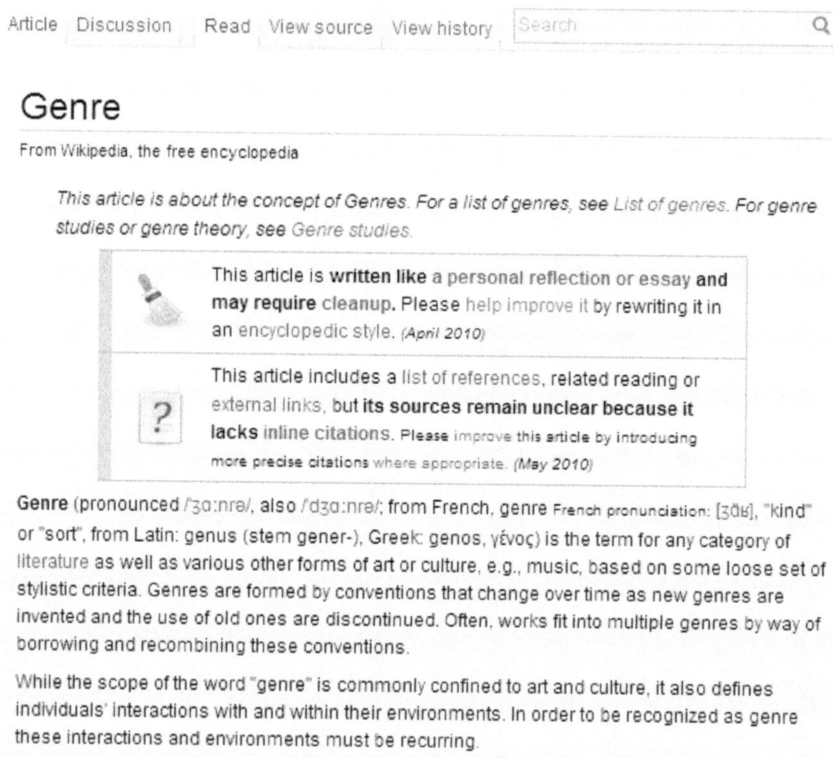
Article Discussion Read View source View history Search

Genre

From Wikipedia, the free encyclopedia

This article is about the concept of Genres. For a list of genres, see List of genres. For genre studies or genre theory, see Genre studies.

This article is **written like** a personal reflection or essay **and may require** cleanup. Please help improve it by rewriting it in an encyclopedic style. *(April 2010)*

This article includes a list of references, related reading or external links, but **its sources remain unclear because it lacks** inline citations. Please improve this article by introducing more precise citations where appropriate. *(May 2010)*

Genre (pronounced /ʒɑːnrə/, also /dʒɑːnrə/; from French, genre French pronunciation: [ʒɑ̃ʁ], "kind" or "sort", from Latin: genus (stem gener-), Greek: genos, γένος) is the term for any category of literature as well as various other forms of art or culture, e.g., music, based on some loose set of stylistic criteria. Genres are formed by conventions that change over time as new genres are invented and the use of old ones are discontinued. Often, works fit into multiple genres by way of borrowing and recombining these conventions.

While the scope of the word "genre" is commonly confined to art and culture, it also defines individuals' interactions with and within their environments. In order to be recognized as genre these interactions and environments must be recurring.

Figure 4. Wikipedia entry for 'Genre' (13/09/2010)

The article is now not dissimilar from the corresponding entry in more traditional encyclopedias, such as for instance *Encyclopedia Britannica*, the online version of which reads:

> **genre**, (French: 'kind' or 'sort') a distinctive type or category of literary composition, such as the epic, tragedy, comedy, novel, and short story. Despite critics' attempts to systematize the art of literature, such categories must retain a degree of flexibility, for they can break down on closer scrutiny. For example, hybrid forms such as the tragicomedy and prose poem are possible. Newly created forms, such as Vikram Seth's *The Golden Gate* (a novel written in

> rhyming verse form) and John Fuller's *Flying to Nowhere* (a novel written in highly poetic prose), and numerous prose works of intermediate or very specific length (such as the novella and the short short) are a clear indication of the difficulty of too close a reliance on genre as a category.

Furthermore the Wikipedia entry has been modified so as to comply with the specific style requirement of an online encyclopedia. For instance, the text has been wikified (i.e. links have been added to other Wikipedia articles) and a link to the pronunciation of the word has been included, thus taking full advantage of the hypertextual environment.

While certainly resembling a traditional encyclopedia entry, this article is nonetheless still considered by the community as inadequate precisely in terms of genre, as the note "This article is written like a personal reflection or essay" by other Wikipedians in their editorial function suggests. What becomes apparent here is the double function of genres that, in Bhatia's words, "on the one hand give suitable expression to the communicative intentions of genre writers" and "on the other hand [they] also match their intentions against their intended reader's expectations", something which can only happen "when all the participants share not only the code but also the knowledge of genre" (Bhatia 2004: 192).

Another evident example of the centrifugal effect of a medium that enables immediate publication of unsupervised content is found in the entry for 'Montezuma'. Here, at one stage of the spontaneous peer-review editorial process, one user commented that "the article is written like a travel guide" and not an encyclopedia entry:

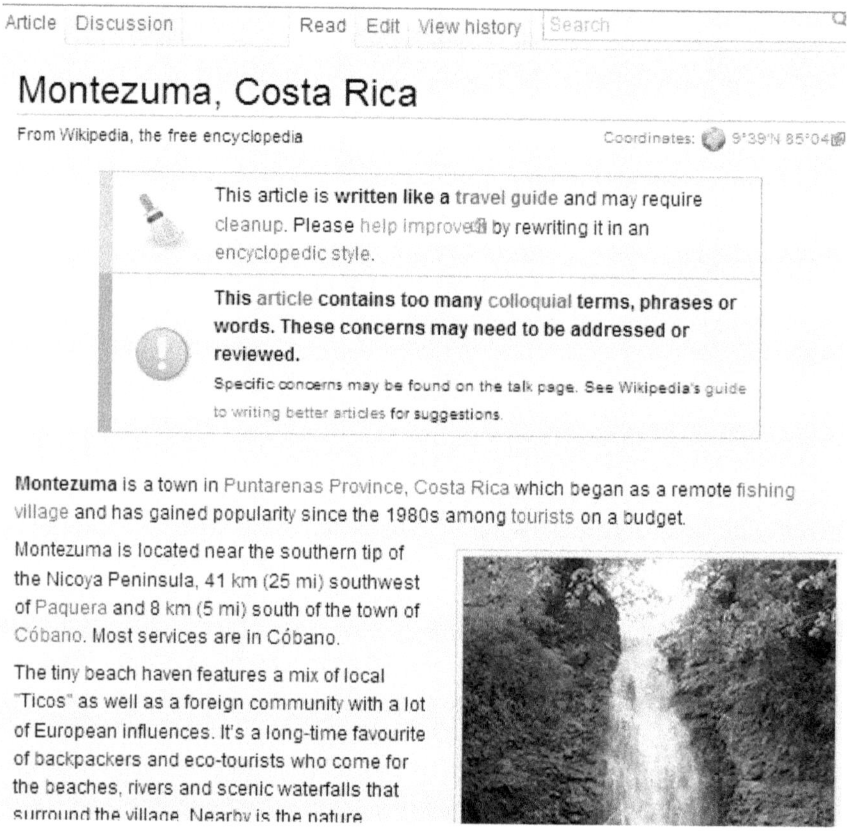

Figure 5. Wikipedia entry for 'Montezuma' (31/07/2010).

The same article is also criticized for containing colloquial terms, phrases and words, all concerns relating to form rather than quality of information content. Similarly, the article on 'Human Genetic Clustering' is labelled as being written "like a scientific review article", again a different genre from the encyclopedia entry:

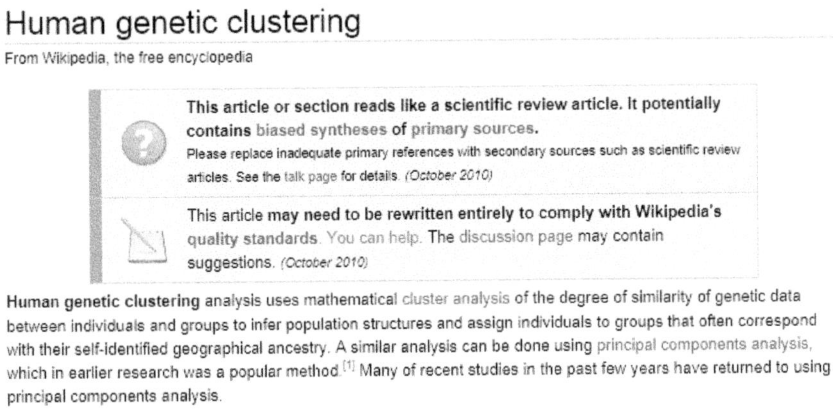

Human genetic clustering

From Wikipedia, the free encyclopedia

This article or section reads like a scientific review article. It potentially contains biased syntheses of primary sources.
Please replace inadequate primary references with secondary sources such as scientific review articles. See the talk page for details. *(October 2010)*

This article **may need to be rewritten entirely to comply with Wikipedia's quality standards**. You can help. The discussion page may contain suggestions. *(October 2010)*

Human genetic clustering analysis uses mathematical cluster analysis of the degree of similarity of genetic data between individuals and groups to infer population structures and assign individuals to groups that often correspond with their self-identified geographical ancestry. A similar analysis can be done using principal components analysis, which in earlier research was a popular method.[1] Many of recent studies in the past few years have returned to using principal components analysis.

Figure 6. Wikipedia entry for 'Human genetic clustering' (25/07/2011).

This concern for the generic integrity of Wikipedia articles is far from episodic. It is on the contrary clearly expressed in a section called 'What Wikipedia is not', which lists other genres or text types that might interfere with the encyclopedic style of Wikipedia:

> Wikipedia is an encyclopedic reference, not an instruction manual, guidebook, or textbook. Wikipedia articles should not read like handbooks or travel guide, or but neither as academic or research papers, text books or journal articles.[4]

What is worth stressing however is that these considerations are not written by an official editorial board, but are rather the outcome of the reactions by ordinary Wikipedians who have from time to time observed a discrepancy between their intuitive appreciation of what makes a 'good' encyclopedia entry and the entries actually produced by users. At the level of the real-life exchange underlying the cooperative writing enterprise of Wikipedia, this has resulted in the decision to make a number of templates available to all users, so as to allow anyone to signal articles which need major clean-up in terms of genre conventions[5]. Thus it is not infrequent to find articles, like those just mentioned, commented on as follows:

4 See <http://en.wikipedia.org/wiki/ Wikipedia:What_Wikipedia_is_not>.

5 See <http://en.wikipedia.org/wiki/Wikipedia:Template_messages/Cleanup>.

> This article is written like a personal reflection or essay rather than an encyclopedic description of the subject.
> This article is written like a manual or guidebook.
> This article is written like a magazine article.
> This article is written like an advertisement.
> This article reads like an obituary.
> This article reads like a textbook.

All this suggests that the generic integrity of Wikipedia cannot be *a priori* taken for granted, but has to be established and preserved *a posteriori* through editorial changes aimed at reducing the centrifugal effect of individual creativity, or lack of generic competence. In this way all users contribute to the creation of a distinctive style of Wikipedia as an online encyclopedia. This style is maintained through a Manual of Style which provides all users – in their multiple functions as authors, readers or editors/reviewers – with detailed guidelines for writing 'good' Wikipedia articles.

4.2. Leads and first sentences

Among the guidelines given in the Manual of Style of Wikipedia, the present section is focused on the guidelines provided for the Lead section of each article, and on the guidelines for writing the first sentence in particular.[6] These parts are recognized by Wiki-pedians as especially important in terms of the pragmatic effect of "establishing the significance" of the entry, and "mak[ing] readers want to know more", and can also contribute to the way the entry is perceived in term of its genre, if we are to believe, as Bakhtin would say, that you recognize a genre "from the very first words" (Bakhtin 1986: 79).

According to the Manual of Style, cooperatively drafted within Wikipedia Community, an article should begin with a short declarative sentence, answering two questions: "What (or who) is the subject?" and "Why is this subject notable?" In the opening section, phrases such as "refers to", "is the name of", or "is a term for" are to be avoided in favour of a concise definition putting the article in con-

6 <http://en.wikipedia.org/wiki/Wikipedia:Manual_of_Style/Lead_section>.

text for the non-specialist. In the history of Wikipedia entries investigated in the present study this is clearly visible in a number of editorial interventions which concern the first sentence of the article. See for instance the entry for 'Atom', as it was conceived at the time of its creation (1/10/2001):

> The ordinary variety of matter that is dealt with in everyday experience consists of discrete units called atoms. The name comes from the Greek atomos, meaning indivisible […].

This was re-edited at a later stage (revision as of 20/09/2002) by choosing 'Atom' as theme and topic for the first sentence and introducing a relational process realized by the verb *to be* as a starting point:

> An **atom** is the smallest, irreducible constituent of a system. Most often the term refers to chemical atoms, the basic constituents of molecules and ordinary matter. These atoms are not divisible by chemical reactions but are now known to be composed of even smaller particles. This article discusses these chemical atoms.

Similarly, in the article 'Green Chemistry Metrics', the first author had clearly no idea of the impact of thematic progression on the accessibility of information, and did not even think of providing a definition:

> Having made a 'Green Chemistry' improvement to a chemical process, it is important to be able to quantify the change. By quantifying the improvement […].

The problem created in this case by the marked theme characterizing the entry's opening sentence at the time of its creation (18/2/2007) is only at a later stage removed by another user who considered it more appropriate to start with an unmarked sentence having 'Green Chemistry Metrics' as subject, and defining it, perhaps a bit tautologically, using the verb *measure*:

> Green chemistry metrics measures efficiency in a chemical process. Having made a Green Chemistry improvement to a chemical process, it is important to be able […]. (revision as of 26/12/2010)

Also revealing are editorial changes in the entry for 'Raman Spectroscopy', where reference to the specific process ("relies on inelastic scattering") which is characteristic of Raman Spectroscopy is deferred to the second sentence, after a more general definition as "a spectroscopic technique used to…":

> Raman spectroscopy relies on inelastic scattering of typically visible light. The near ultraviolet and near infrared are also used. Inelastic scattering is tpyically [sic] very weak. Therefore the technique only became feasible and widely used after lasers became available. (First version: 07/08/2002)

> Raman spectroscopy (/ˈrɑːmən/; named after C. V. Raman) is a spectroscopic technique used to study vibrational, rotational, and other low-frequency modes in a system.[1] It relies on inelastic scattering, or Raman scattering, of monochromatic light, usually from a laser in the visible, near infrared, or near ultraviolet range. (Revision as of 28/09/2010)

Similarly, the entry for 'Sandwich compound' as of 25/07/2006 read as:

> The term sandwich compound was introduced in organometallic nomenclature in during the mid-1950s when J.D. Dunitz, L.E. Orgel and R.A. Rich discovered from X-ray crystallography the particular structure of ferrocene and their conformers which exhibited particular [sic] an iron atom sandwiched between two parallel cyclopentane rings. It is said that the subsecuent [sic] studies after it, began the era of modern organometallic chemistry.

and was re-edited as follows:

> In organometallic chemistry, a sandwich compound is a chemical compound featuring a metal bound by haptic covalent bonds to two arene ligands. The arenes have the formula CnHn, substituted derivatives (for example Cn(CH3)n) and heterocyclic derivatives (for example BCnHn+1). Because the metal is usually situated between the two rings, it is said to be 'sandwiched.' A special class of sandwich complexes are the metallocenes. (revision as of 9/08/2011)

Here, again, the definition "a sandwich compound is a chemical compound…" has been opted for as a more adequate opening sentence, using a marked theme, "In organometallic chemistry", as a disambiguator.

Changes of this kind are relatively frequent, and a preliminary quantitative analysis based on a corpus of Wikipedia entries on Chemistry suggests that the tendency to open the article with the pattern "X is a/an/the", where X is often the same as the title of the article, is characteristic of later stages in the evolution of the Wikipedia articles contained in the corpus. Comparing data from the 160,000 token WikiChem corpus composed of Wikipedia articles on Chemistry, with data from the smaller WikiChem1 corpus made up of the corresponding earliest version for each of the articles included in WikiChem suggests that the patterns "is a/an" and "is the" occur less frequently in the early versions, and only in 50% of these articles are these patterns found in the first sentence. This percentage rises instead to 87% in the corpus including only late entries. All this seems to suggest that, besides relating to the quality of content in Wikipedia articles, or to their functionality (e.g. through Wiki-fication), some changes definitely relate to the language used and specifically to the articles' adherence to the encyclopedic style of Wikipedia in ways that suggest a strong concern for its generic integrity. The result is a digital encyclopedia embodying a hybrid genre that combines the best that fixity and flexibility can give.

5. Conclusion

In this chapter the contribution of the technological affordances of Web 2.0 to the action of the centripetal and centrifugal forces of discourse (Bakhtin 1982) has been investigated with specific reference to the case of Wikipedia, the well-known cooperative online encyclopaedia. In particular, investigating the changes occurring in the development of single entries, by contrasting early and later stages, seemed to provide a viable way to find evidence of the interaction of the centrifugal forces unleashed by the technological empowerment of each single individual as a potential author, and the centripetal effect of *post-hoc* revision and editorial interventions by a community of readers/authors/editors. Preliminary findings from a sample of entries indicate that a more extensive quantitative/qualitative analysis of data from the History of Wikipedia articles might shed new light on the patterns of negotiation and potential conflict through which generic integrity is achieved and preserved in this prototypical Web 2.0 genre.

References

Askehave, Inger / Ellerup Nielsen, Anne 2005. What are the Characteristics of Digital Genres? Genre Theory from a Multi-Modal Perspective. *Proceedings of the 38th Annual Hawaii International Conference on System Sciences (HICSS'05).*

Bakhtin, Mikhail Mikhaĭlovich 1982. *The Dialogic Imagination. Four Essays by M.M. Bakhtin.* Austin: University of Texas Press.

Bakhtin, Mikhail Mikhaĭlovich 1986. *Speech Genres and Other Late Essays.* Austin: University of Texas Press.

Bateman, John 2008. *Multimodality and Genre: A Foundation for the Systematic Analysis of Multimodal Documents.* Basingstoke: Palgrave Macmillan.

Berkenkotter, Carol / Huckin, Thomas N. 1995. *Genre Knowledge in Disciplinary Communication: Cognition/Culture/Power.* Hillsdale, NJ: Lawrence Erlbaum.

Bhatia, Vijay Kumar 1997. The Power and Politics of Genre. *World Englishes.* 16/3, 359-371.

Bhatia, Vijay Kumar 2004. *Worlds of Written Discourse.* London: Continuum.

Bruns, Axel 2008. *Blogs, Wikipedia, Second life, and Beyond: From Production to Produsage.* New York: Peter Lang.

Crawford, Holly 2001. Encyclopedias. In Bopp, Richard E. / Smith, Linda C. (eds), *Reference and Information Services: An Introduction.* Englewood, CO: Libraries Unlimited, 433-459.

Cusinato, Alberto / Di Francesco, Salvatore / Della Mea, Vincenzo / Mizzaro, Stefano 2009. QuWi. Quality Control in Wikipedia. *Proceedings of WICOW09*, April 20, 2009, Madrid, Spain.

Eisner, Caroline / Vicinus, Martha (eds) 2008. *Originality, Imitation, and Plagiarism: Teaching Writing in the Digital Age*. Ann Arbor: University of Michigan Press.

Giltrow, Janet / Stein, Dieter (eds) 2009. *Genres in the Internet: Issues in the Theory of Genre*. Amsterdam: Benjamins.

Kelly, Kevin 2005. We are the Web. *Wired*, 13th August 2005 <http:// www.wired.com/wired/archive/13.08/tech.html>.

Miller, Carolyn R. / Shepherd Dawn 2010. Questions for genre theory from the blogosphere. In Giltrow, Janet / Stein, Dieter (eds) *Genres in the Internet: Issues in the Theory of* Genre. Amsterdam: Benjamins, 263-290.

Peters, Isabella 2009. *Folksonomies: Indexing and Retrieval in Web 2.0.* Berlin: Walter de Gruyter, 2009.

Ray, Amit / Graeff, Erhardt 2008. Reviewing the Author-Function in the Age of Wikipedia. In Eisner, Caroline / Vicinus, Martha. *Originality, Imitation, and Plagiarism: Teaching Writing in the Digital Age*. Ann Arbor: University of Michigan Press, 39-47.

Santini, Marina 2005. Web pages, text types and linguistic features: some issues. *ICAME Journal*. 30, 67- 86.

Santini, Marina 2007. Characterizing Genres of Web Pages: Genre Hybridism and Individualization. *40th Annual Hawaii International Conference on System Sciences (HICSS'07)*. <http:// csdl2.computer.org/comp/proceedings/hicss/2007/2755/00/2755 0071.pdf>.

Santini, Marina / Mehler, Alexander / Sharoff, Serge 2010. Riding the Rough Waves of Genre on the Web. In Mehler, Alexander/ Sharoff, Serge / Santini, Marina. *Genres on the Web: Computational Models and Empirical Studies*, 3-32.

Stvilia, Besiki / Twidale, Michael B. / Gasser, Les / Smith, Linda C. 2005. Information Quality Discussions in Wikipedia. *Proceedings of ICKM05*.

Toffler, Alvin 1981. *The Third Wave*. New York: Bantam Books.

Wesch, Michael 2007. *Web 2.0: The machine is us/ing us* [Video]. <http: //www.youtube.com/watch?v=6gmP4nk0EOE>.

Wikipedia: Featured Articles, <http://en.wikipedia.org/wiki/Wikipedia:Featured_articles>.

Wikipedia: What Wikipedia is not, <http://en.wikipedia.org/wiki/Wikipedia:What_Wikipedia_is_not>.

Wikipedia: Manual of Style, <http://en.wikipedia.org/wiki/Wikipedia:Manual_of_Style>.

Yates, Simeon J. / Sumner, Tamara R. 1997. Digital Genres and the New Burden of Fixity. In *Proceedings of the 30th Hawaiian International Conference on System Sciences (HICCS 30).* VI, 3 - 12.

ENRICO GRAZZI

The Web as a Participatory Environment: Social Networks and 'Memes' from a Teacher's Perspective

The aim of this chapter is to study the use of English as a *lingua franca* (ELF) on the Internet and focus on the spread of memes as an example of the creative innovations introduced by the impact of social networks on informal discourse. Since the advent of Web 2.0, online interaction and intercultural communication via ELF have grown globally, leading to the creation of cyber communities, whose members are drawn together by shared interests and common goals. Following Lantolf and Thorne (2006: 67), it is claimed that the World Wide Web should be included in the list of settings where developmental processes take place, hence it would seem appropriate to consider ELF as a cultural artifact that allows 'Internet-mediated communication'. However, considering ELF speakers as members of a community is a highly controversial issue, especially if we take into consideration the *double identity* of most non-native speakers (NNSs) of English, who are both EFL learners (i.e. students living in countries where English is not the L1, and who are attending or used to attend institutional language courses at school or university), and ELF users (when they use English as a contact language within authentic communicative contexts, e.g. the Web). The conflicting ideas behind the academic debate on the nature of EFL (*English as a Foreign Language*), ELF and the NNSs' identity seem to leave no space for compromise. Nonetheless, it is suggested that a Vygotskyan approach to language learning may reconcile these opposing views and provide the appropriate key to interpret complex sociolinguistic phenomena such as those referred to here.

1. ELF, EFL and SCT:[1] an integrated approach

This chapter is part of an ongoing research project called: "The convergence of English as a Lingua Franca and English as a Foreign Language, and its pedagogical implications", which is being carried out at the Dept. of Linguistics of the University of Roma Tre. The fundamental assumption of this research is that the Internet should be considered an authentic social environment, inhabited by culturally and ethnically diverse communities who mainly use ELF to interact online. In a globalized world, the heterogeneity of the growing population of net users is perhaps the most tangible manifestation of cross-cultural communication, and the pervasiveness of English as the primary contact language has given a strong impulse to the emergence and swift spread of its non-standard varieties, a phenomenon that can be interpreted as the inevitable consequence of the NNSs' tendency to adapt the international language to their own linguistic and cultural identity, in order to fulfil their communication needs. As Crystal (1997: 137) noted, people tend to use a distinctive accent and dialect to mark their own nationality on the world stage:

> These differences become especially noticeable in informal settings; for example they are currently well represented in discussion groups on the Internet. [...] International varieties thus express national identities, and are a way of reducing the conflict between intelligibility and identity.

The nature of ELF, however, is still a matter of debate amongst linguists, whose opposing views and orientations might essentially be grouped into three prevailing attitudes, which are summarized here. The first one shows disparagement for any deviant form of English spoken by the NNSs of the Expanding Circle, and reaffirms the unifying role of official standard Englishes. The second attitude shows the acceptance of Expanding Circle Englishes (Kachru 1986) as legitimate local dialects, although a new standard form of spoken English, strongly influenced by US English, is envisaged to allow mutual understanding in international settings and avoid the risk of fragmenta-

---

1 SocioCultural Theory.

tion. Finally, the third attitude recognizes ELF as a language in its own right, enriched by the sociocultural identities of its speakers. ELF, it is claimed, has arisen naturally from the contact of native-speaker (NS) varieties of English and NNS mother tongues, in the attempt to carry out international communication. Once started, this process cannot be stopped and should not be submitted to extraneous restrictions and norms.[2]

As regards the scientific background of this research, I have essentially sided with those who support this third position, although a subtle distinction has to be made at this point, to give a more accurate description of the theoretical framework of this study. Since Kachru's tripartite classification of English-speaking countries into 'Inner Circle', 'Outer Circle' and 'Expanding Circle' has become a widely accepted model, there has been a general tendency to use these definitions not only to represent the geographical, cultural and historical reasons behind this subdivision, but also to refer to the different communities of speakers of English who belong to these three areas. Nevertheless, this schematization, however practical it may be, oversimplifies a far more complex sociolinguistic reality, especially as regards the Expanding Circle, which by definition is characterized by social diversity and heterogeneity. The controversial point is that NNSs of English can hardly be considered a *community* in a general sense, and the fact that most of them study or used to study English as a foreign language in their home countries and use this language as a *lingua franca* for practical purposes does not automatically grant shared community membership to populations that do not constitute a proper speech community.[3]

2 See Jenkins (2000: 30-32) for a concise summary of the different positions on interlanguage competence.

3 Although there is not a unitary definition of speech community, most sociolinguists consider shared community membership and shared linguistic communication as typical features of speech communities. According to Morgan (2004: 4): "The concept of speech community does not simply focus on groups that speak the same language. Rather, the concept takes as fact that language represents, embodies, constructs and constitutes meaningful participation in a society and culture. [...] Thus it is within the speech community that identity, ideology, and agency [...] are actualized in society. [...] [M]embership in a speech community includes local knowledge of the way

This leads to the logical conclusion that NNSs cannot be considered a single social body, but rather a set of very diverse communities, and in turn ELF should not be considered as a language in the conventional sense, because it lacks a specific sociocultural background. The uniqueness of ELF consists in the fact that, while conventionally there is supposed to be a one-to-one relationship between a language and its community of speakers, here we have a one-to-many relationship, which results in the proliferation of novel NNS varieties of English (e.g. Euro-English, Asian Englishes, etc.). Besides, according to Prodromou (2008: xiv), Kachru's diagram of the three concentric circles misrepresents the real cross-cultural nature of ELF, because the fact that these circles delimit three distinct areas might lead to the false assumption that ELF is a prerogative of NNSs of English, i.e. of the expanding circle. On the contrary, the three circles should rather be represented as intersecting socio-linguistic areas, where ELF emerges as the "outcome of all circles interacting". ELF, in other words, is seen as a tool for negotiation across these three circles, and as regards the rise of new local varieties of English, Prodromou (2008: xv) adds:

> The process of diversification does not necessarily involve rejecting the forms of Standard English but adding to them on a lexico-semantic level. It is a process of appropriation of discourse repertoires which may be expressed, in some cases, by *simplification*, but also by *amplification*.

Anyhow, it seems quite reasonable to think that there is still an unresolved puzzle at the heart of the conflict between those who advocate the legitimacy of considering ELF on a par with other languages, and those who, on the contrary, see non-native varieties of English as unstable interlanguages[4] that gradually fos-

---

language choice, variation, and discourse represents generation, occupation, politics, social relationships, identity, etc."

4 Selinker (1972) defined the language learner's language as *interlanguage* (IL), that is a language system that develops between the two extremes of the student's first language (L1) and the Target Language (TL). According to Pit Corder (1981: 17) the IL is an unstable idiosyncratic dialect whose "conventions are not shared by a social group [and] many of its sentences present problems of interpretation to any native speaker of the target language. [...] It

silize[5] and turn into a sort of broken English.[6] Both parties seem to have sound reasons underpinning their views, but the cause of the ambiguity surrounding this dispute has to be ascribed to the identity of the L2-user[7] being suspended between the global citizen and the language learner. To disambiguate their twofold nature, we would rather place it in a fresh perspective, for in reality it represents the two sides of the same coin. The polarity of ELF and Interlanguage marks the fundamental distinction between the pragmatic use of English to communicate in authentic international contexts, and the process of teaching/learning English as a foreign language (EFL) in an institutional context (e.g. at school). While these two domains are interrelated and convergent, as will be shown later on, they should not be automatically conflated in approaching the phenomena that characterize ELF and EFL. Hence, even though the figures of the L2-user and the language learner usually coincide, they are not interchangeable and it is important to distinguish between different foci when we consider the circumstances and variables that determine the emergence of ELF[8] on the

---

is a dialect whose rules share characteristics of two social dialects of languages, whether these languages themselves share rules or not."

5 According to Selinker (1972: 215): "Fossilizable linguistic phenomena are linguistic items, rules, and subsystems which speakers of a particular native language will tend to keep in their interlanguage relative to a particular target language, no matter what the age of the learner or amount of explanation or instruction he receives in the target language."

6 Quirk (1988, in Seidlhofer 2003: 12), for example, says that: "The problem with varieties in this branch is that they are inherently unstable, ranged along a qualitative cline, with each speaker seeking to move to a point where the varietal characteristics reach vanishing point, and where thus, ironically, each variety is best manifest in those who by commonsense measures speak it worst."

7 Following Cook (2002: 4), I refer to the NNS of English as an L2-user, that is: "A person who uses another language for any purpose at whatever level." Therefore, the expression L2-user is an all-inclusive term that refers also to the NNS as an L2-learner.

8 Among these variables and circumstances, contact-induced language change plays an important role. Quoting Heine and Kuteva (2005: 34-37): "Rather than viewing replication as leading to a 'deviation from the norm' or as a disruption of an existing state, we view it as leading to a new state that is simply different from the earlier state, but is not necessarily less coherent, less 'sys-

one hand, and the pedagogical components that contribute to the student's transitional competence[9] on the other. In other words, we should use the expression ELF when we take into consideration the L2-user's authentic communicative performance, and conversely EFL or IL when we deal with the non-native student's learning process. As Jenkins (2000: 10-11) points out:

> One possible alternative to EFL simply involves reversing the second and third letters to arrive at ELF, or 'English as a Lingua Franca'. This term would have a number of immediate advantages: ELF emphasizes the role of English in communication between speakers from different L1s, i.e. the primary reason for learning English today; *it suggests the idea of community* as opposed to alienness; it emphasizes that people have something in common rather than their differences; it implies that mixing languages is acceptable [...] and thus that there is nothing inherently wrong in retaining certain characteristics of the L1, such as accent; finally, the Latin name symbolically removes the ownership of English from the Anglos both to no one and, in effect, to everyone. [emphasis added]

---

tematic', or less complete than the earlier state of the language concerned. And rather than viewing speakers as receivers, imperfect language learners, etc., we find massive evidence for a perspective according to which speakers are more appropriately analyzed as actors and 'language builders' [They are] creators who use what they find in one language and sociocultural environment to shape another language in novel ways – they do not simply imitate grammatical categories, or produce imperfect copies of such categories; rather, they are likely to develop new use patterns and new categories on the model of other languages."

9 The term *transitional competence* was originally used by Pit Corder (1981: 10, 34) to define the learning continuum that is typical of interlanguage: "What he [the learner] speaks is not an inadequate or incorrect form of the target language but a peculiar transitional idiolect". Moreover, Pit Corder (1981: 93) suggests that: "Language is multidimensional, consisting of two basic kinds of continua – developmental, characterized by increasing complexity towards some particular target in the case of a language learner, and lectal (including dialect chains which also connect so-called different languages) characterized by equal complexity but oriented towards some reference-norm – in the case of any particular language community. [...] My proposal now is that language learning, certainly in the case of a first language, is a creative process and yields in the individual a purely developmental continuum. In practice,–second language learning in any particular individual is probably a mixture in varying proportions of restructuring and recreating."

What seems to be less persuasive in Jenkins's position is the idea of ELF speakers as de facto members of a community, for the reasons explained above. Presumably, what Jenkins is implying here is the tremendous importance of the sociocultural change brought about by a burgeoning common language that connects millions of people around the world. I subscribe to this view and have therefore chosen to focus on the use of ELF on the Internet, which, as mentioned at the beginning of this chapter, is an authentic environment where international communication takes place. I would rather use the term *cyber community* (CC) in this case, which suits the population of net users and social networkers, who usually gather in peer subgroups, drawn together by shared interests and ideas, as well as their expertise and common goals. To some extent, CC groups could be likened to Wenger's (1998: 78) *communities of practice*, a term from SCT that has been successfully used also in linguistic contexts. As Wenger himself explains on his personal website (<http://www.ewenger.com/ theory/>):

> The concept of community of practice is influencing theory and practice in many domains. [...] New technologies such as the Internet have extended the reach of our interactions beyond the geographical limitations of traditional communities, but the increase in flow of information does not obviate the need for community. In fact, it expands the possibilities for community and calls for new kinds of communities based on shared practice.

The idea of a web community as a learning environment based on shared know-how suggests that there are interesting pedagogical implications of this theory for ELF, as regards English language teaching (ELT). Therefore, it is advisable that the sociolinguistic component (which has been relatively neglected in the controversial debate around the nature of ELF) should be taken into account when the focus of research is shifted to the pragmatic use of English as an international language. Even the most notable examples of applied research in the field of ELF, such as Jenkins's (2000: 123-163), attempt to establish a phonological core for ELF, or Seidlhofer's (2004: 209-239) compilation of VOICE 1.0 (the first free-access corpus comprising a million lexicogrammatical items of spoken ELF), are essentially concerned with contrastive studies at language level, while a systematic study of ELF that integrated applied linguistics and the sociocultural approach has very infrequently been undertaken.[10] This should incorporate Vygotsky's (1962) Sociocultural Theory (SCT) and the recent work by scholars such as Lantolf and Thorne (Lantolf 2000; Lantolf/Thorne 2006), whose approach to second language learning provides a theoretical framework that is applicable to the study of the complex phenomena characterizing online communication and the concurrent evolution of ELF as a global means of mediation.[11] According to Lantolf and Thorne (2007: 201):

> Developmental processes take place through participation in cultural, linguistic, and historically formed settings such as family life and peer group interaction, and in institutional contexts like schooling, organized sports activities, and work places, to name only a few.

Hence, by claiming that the World Wide Web should be included in the list of settings where developmental processes take place, it may be feasible to study ELF as a flourishing cultural artifact devised to cope with what Lantolf and Thorne (2006: 67) have termed 'Internet-mediated communication'. Moreover, the fact that CC members who interact through ELF are often EFL learners who study English in school or university makes the Internet a valuable resource for language teachers and students. This entails the convergence of ELF and EFL and the possibility of observing interesting phenomena in the evolution of the CC's lexicogrammatical system, which may be more deeply investigated by integrating the SCT approach and SLA theories. The creation of memes by social networkers is a case in point, as will be shown in the following section, and it represents a first step for further research in this area.

---

10 A case in point is Prodromou's (2008) study on ELF, which draws on Vygostky, Lantolf and Thorne's socio-cultural approach, as well as on Bakhtin's cultural theory.

11 Lantolf and Thorne's approach to second language development is informed by Vygotsky's sociocultural theory. This chapter draws essentially on their research for what concerns the implementation of Vygotskyan cultural-historical psychology in second language acquisition (SLA) research.

2. A case study: the creation of memes

The advent of Web 2.0 has marked a great change for the World Wide Web, as this new technology has turned the Internet into a dynamic system that allows people to be connected, upload and exchange user-generated contents, and interact online. Social networks like MySpace, Facebook, and Twitter have allowed the creation of online communities, providing group-centred services for people who come together for different reasons, e.g. to share views and ideas on various topics, to keep in touch with friends and relatives, and to cooperate in areas of common interest. ELF has become the prevalent language for international communication both within CC groups and virtual communities of practice, thus supplying a potentially enormous amount of relevant data, which pave the way for further research in the fields of applied linguistics and ELT. The particular intent of this study is to focus on the creation and the use of Internet memes[12] that combine pictures with verbal messages, a unique language phenomenon that is peculiar to informal online communication among social networkers. This kind of Internet memes consists in the unprecedented combination of an emblematic visual image (usually a weird and funny photo available online) that portrays an unusual situation, and a short verbal text (usually a single-word clause or an acronym that stands for an

12 In 1976, the word *meme* was introduced by the British scientist R. Dawkins (1989: 186), who explains its origin as follows: "We need a name for the new replicator, a noun that conveys the idea of a unit of cultural transmission, or a unit of imitation. 'Mimeme' comes from a suitable Greek root, but I want a monosyllable that sounds a bit like 'gene'. I hope my classicist friends will forgive me if I abbreviate mimeme to meme. If it is any consolation, it could alternatively be thought of as being related to 'memory', or to the French word *même*. It should be pronounced to rhyme with 'cream'." By analogy with Dawkins's meme, an Internet meme is an idea that spreads swiftly through the Web in the form of a hyperlink, a video, a picture, a word or phrase, etc. Internet memes usually have a strong public impact. They are shared through social networks, blogs, e-mail messages, etc. and net users appropriate and often adapt them to fit their communicative needs. Because of their great popularity, internet memes have also been successfully used in marketing as a form of low-cost advertising.

idiom, an exclamation or an interjection) associated to it. Similarly to advertisements, which connect a picture to a payoff line to create a verbo-pictorial metaphor,[13] the iconicity of memes makes their messages extremely evocative and effective, and their intent is to induce an emotional response from the receiver.

Typically, memes are ironic and are found as part of discourse in forums and blogs, that is when participants are not involved in instant messaging and therefore have more time to create their original memes or choose among the existing ones. Through memes, language becomes more emphatic and expresses connotative meanings which the members of a given community are able to decode. Their metaphorical structure, in fact, combines images and verbal texts according to a semiotic process based on symbolic exophoric references to given situational contexts and cultural elements. This is also the reason why in most cases the new members of a CC group or subgroup find it hard to get the messages conveyed by memes, for they still lack the necessary background knowledge to infer their hinted meanings. Internet memes, like genes in organisms, replicate each time they are passed on from one social networker to another, so they spread very quickly within a certain group. Besides, as group members are often members of other groups, memes are easily exported to other situational contexts and become part of a wider shared code, till it becomes almost impossible to trace them back to their origins. Figures 1 and 2 are typical examples of memes.[14]

13 Quoting Forceville (1996), Lombardo (1999) defines verbo-pictorial metaphors as those in which one of the subjects is pictured while the other is in the verbal text.

14 The memes that are presented in this chapter were downloaded from the Internet. Nobody owns their copyright, and they are normally exchanged and modified by anonymous social networkers.

Figure 1.

Figure 2.

The meme in Fig. 1 is used ironically to express disagreement or disappointment for something that was not done properly or that went wrong, in other words, a failure. This is metaphorically represented by the mismatch between the sign and the dresses on the stand, and is reinforced by the word 'Fail'. In Fig. 2, instead, we have the funny picture of a bird, and the acronym 'O RLY?' that stands for the question: 'Oh, really?' This meme is used to express surprise and disbelief. Interestingly, the answer to this meme can be a similar meme, but with a completely different meaning, as in Fig. 3. Here, the combination of a picture of a frowning owl and the acronym 'YA RLY' (that stands for the answer: 'Yes, really') well expresses the illocutionary force of a definitive answer.

Figure 3.

Other common acronyms we find in Internet memes are for example: OMG (Oh, my God!); BTW (By the way); LOL (Laughing out loud) (Fig. 4).

Figure 4.

Word play is also very common as a form of the creative language of memes. The next example (Fig. 5) shows the use of a blend in the verbal part of the message, whereby two partial homophones, *more* and *roar*, form a new word, *moar*, which is associated to the photo of a hungry cat, as if it were actually roaring to say: "I'm hungry! Give me some more food!" Social networkers use this meme especially when they are eager to know more about something they are discussing online.

Figure 5.

The comparison between Internet memes and genes could be expanded to include another fundamental characteristic of genetic duplication, that is random mutations. It has been observed that the spread of memes across the Internet (the 'memic pool', to use Dawkins's words) is based on a process of imitation, which is often the cause of voluntary or involuntary changes and alterations. These may consist in misspelt words and acronyms, or even typing mistakes that are not going to be corrected, but on the contrary are kept and replicated, and soon become a recognizable feature of certain memes. We could even say that the intentional use of *deviant* memes marks the fact that the user belongs to a certain CC group or subgroup, thus they take on an important sociocultural function as far as they become a sign of the social networker's identity. Figure 6 shows a popular 'deviant' meme.

Figure 6.

Here we have a double typing mistake for the word 'owned'. Presumably, the addition of the initial -p and the replacement of *3* instead of *e* were due to the fact that the keys for *o* and *p* are very close on the computer keyboard, as well as the keys for *e* and *3*, so we could assume that the typist may have pressed the wrong keys accidentally. The intended informal idiomatic expression is: 'Owned!', meaning at least two different things: "I have you in my power!" (e.g. when someone beats a rival player at a videogame), and: "How humiliating!" (e.g. when someone has suffered a humiliation). This expression has become very popular on the Internet thanks to a videogame called World of Warcraft, where it appears whenever a player is eliminated (e.g. "Player # has been owned"). There are several hypotheses on the origin of the misspelt word *pown3d* (for example, some say it was written by mistake on the map used to play Warcraft), however, it was immediately borrowed by social networkers who were also videogame players, and soon spread across the Internet.

3. The use of memes and ELT

My research project on the convergence of ELF and EFL has taken into consideration the pedagogical implications of using ELF at school to allow Internet-mediated communication. Hence, because memes have become a typical feature of the informal language used by social networkers (especially younger netizens), English teachers could exploit this resource to make their students[15] carry out innovative language activities, whose aim is to turn them into members of a wider CC.[16] Tasks should be graded according to the pupils' age range and language competencies, but it is also possible to assign different tasks (i.e. easier or more complex ones) to different groups of learners within the same class by taking into consideration their different abilities. For instance, the teacher could plan a series of lessons on social networking, where students are allowed to use the school computer room to access the Web and join international discussion groups, blogs or forums dedicated to topics of their interest (e.g. music, sports, books, films, manga, fashion, etc.). Once they are interacting online, they should pay attention to the memes that are used by other networkers and utilize them appropriately in their turn. Besides, students should feel free to personalize these memes and make them suitable to the conversations they take part in, for example by choosing fresh combinations of pictures and verbal messages. Memes, as shown in the previous paragraph, exploit the expressive function of language in online communication and stimulate the L2-user's creativity in a playful mood that contributes to developing a closer rapport with their peers. Therefore I believe that they could help learners to lay the

15 I hereby refer to high school students, whose English level is usually supposed to range between lower-intermediate and upper intermediate.

16 The communicative activities based on the creation of internet memes that I am describing in this section were implemented by a group of Italian high-school teachers of English who attended an in-service training workshop that I ran in 2009, on behalf of TESOL-Italy (a qualified non-profit association of English teachers that is an affiliate of TESOL International Association) and the *Direzione Scolastica Regionale* (Regional Educational Board) of Latium, Rome, Italy.

foundation stone of an open public space for authentic communication that extends beyond the classroom, to the world outside.

Let us now take into consideration the possible disadvantages of introducing social networking into the English syllabus. For example, Goldman *et al.* (2008: 185) say that nowadays more and more students, especially at high school level, are already expert social networkers, hence some might observe that they do not really need the school to provide them with any training in this particular field. Besides, students could even object to the idea that their teacher and classmates should pry into their personal online messaging, as it would be a violation of their privacy.

Although these seem like reasonable observations, one should also consider two important points: a) social networking, at least in Italy, is normally carried out in the users' L1 (i.e. standard Italian, regional dialects, jargons, etc.) to maintain and reinforce their friendship bonds within a given discourse community. Using ELF as a contact language for cross-cultural communication would offer an opportunity to get access to an international community of social networkers, which students would very probably welcome and appreciate; b) through guided activities learners would gradually become more confident ELF-users and benefit from their teachers' assistance as well as from peer-interaction. This leads us to consider the incorporation of social networking within the EFL syllabus from a Vygotskyan perspective, whereby English is intended as a mediational tool whose social and educational affordances allow L2-users to carry out authentic communicative activities, and meanwhile take advantage of the feedback offered by their teachers and peers. Consequently, the practice of social networking and the students' engagement in a meaningful communicative performance is surmised to generate a zone of proximal development (ZPD)[17] that enhances the learner's skills and competencies. Therefore, it is possible to say that within the social ecology of the EFL class, ELF reveals its dual nature, as it acts both as

17 Vygotsky (1978: 86) defined the ZPD as: "The distance between the actual developmental level as determined by independent problem solving and the level of potential development as determined through problem solving under adult guidance, or in collaboration with more capable peers".

a pragmatic tool emerging in intercultural web-mediated activity, and as a catalyst for culturally mediated language development.

The connection of the ZPD concept to ELT and the use of ELF in social-networking entails a redefinition of the roles of the language teacher and the student's peers, because after having bridged the gap between the use of EFL in schools and the authentic use of ELF online, these figures interact within a wider social environment that is open to the participation of all the members of the CC. In this context, the language teacher should essentially behave as a tutor and counsellor who, as Walqui and van Lier (2010: 12) say, provides:

> 'scaffolding' [...] that enable[s] students to make optimal learning gains. [...] The theoretical intent of scaffolding [is] to create the contexts and supports that allow students to interact in their zone of proximal development.

Moreover, when students are involved in web-mediated communication at school, the teacher should not only respond to the student's requests for language counseling and technical support, but also provide appropriate indications for a safe use of social networking, especially to protect teenagers' digital identity. This is the case with sensitive personal information and details that learners could unwisely give out online (e.g. about religion, political views, sexuality, etc.). However, we should also consider that the role of the teacher as educator is not to prevent any potential mistake arising from the student's failure to perceive the risks connected to social networking, but rather to offer advice and guidance whenever it is necessary. As Boyd (2008: 137) suggests:

> By allowing us to have a collective experience with people who are both like and unlike us, public life validates the reality that we are experiencing. We are doing our youth a disservice if we believe that we can protect them from the world by limiting their access to public life. They must enter that arena, make mistakes, and learn from them. Our role as adults is not to be their policemen, but to be their guide.

As for the role of the student's peers, their collaborative participation in the dialogic interaction online activates a process of co-construction and negotiation of meaning among individuals from different cultures, which is expected to lead to reciprocal understanding and language improvements. As Ohta (2001: 76, quoted in Lantolf/Thorne 2006: 284) observes, "when learners work together [...] strengths and weaknesses may be pooled, creating a greater expertise for the group than any of the individuals involved". So, with the implementation of digital media and social networking at school via the use of ELF, the sharing of such expertise ought to be considered an opportunity to focus the learners' attention on language awareness, media literacy and a participatory approach to cross-cultural communication. This however, should not be expected to be a quick process that produces immediate results, but rather a long term one, as Goldman *et al.* have shown (2008: 194):

> Learning doesn't necessarily happen instantly, but develops over time as youth have the structured opportunity to reflect on and critique what they have done, with the goal of improving the next time around. 'Social technologies' give digital media purpose as a means for youth to craft new identities as learners.

4. Conclusions

This study has approached the convergence of ELF and EFL from a sociocultural point of view, taking into consideration the meaning of the word *community* when referring to L2-users of English. Having considered the Internet as an authentic environment where cross-cultural communication takes place, I have focused on international social networking, and particularly on the use of memes, assuming that it is possible to find similarities between the creation and the spread of memes worldwide – even containing deviant language forms – and the emergence of the distinct lexicogrammar system of ELF.[18]

---

18 Notwithstanding the differences between memes and ELF, what I would like to highlight is the similarity between the way memes are spontaneously created and appropriated by social networkers, and the way innovative ELF lexicogrammar forms develop and spread across the community of L2-users. This process and the fact that memes often contain deviant lexicogrammar forms is in my opinion what links them to ELF.

As regards the pedagogical implications of this phenomenon, it is advisable that teachers of English implement digital technology with adolescent-age students and engage them in online communication activities (e.g. through discussion groups, forums, blogs, etc.) in order to bridge the gap between the artificial use of EFL within the institutional context of the classroom and the authentic use of ELF in the international context of the cyber community. This entails a new definition of the roles of the teacher, the learners and the learners' peers, whose interconnected actions generate a ZPD that is crucial to successful learning and to the development of the student's identity. Quoting Goldman *et al.* (2008: 202), we can say that:

> Identity development is an emergent process that comes with sustained participation in existing community structures and discourses. Adults in these cases are intentionally guiding youth through learning processes that support them in intervening in the communities in which they seek to participate.

In conclusion, the findings of the present study indicate that EFL and ELF are not incompatible and that through guided social-networking activities it is not only possible to enhance the learners' communicative competence, but also to further their cross-cultural competence and media literacy.

References

Boyd, Danah 2008. Why Youth ♥ Social Network Sites: The Role of Networked Publics in Teenage Social Life. In Buckingham D. (ed.) *Youth, Identity, and Digital Media*, MacArthur Foundation Series on Digital Learning, Cambridge, MA: MIT Press, 119-142.

Cook, Vivian (ed.) 2002. *Portraits of the L2 User*. Clevedon, Multilingual Matters.

Crystal, David 1997. *English as a Global Language*. Cambridge: Cambridge University Press.

Dawkins, Richard 1989. *The Selfish Gene* (2nd ed.). Oxford: Oxford University Press.

Forceville, Charles 1996. *Pictorial Metaphor in Advertising*. London/New York: Routledge.

Goldman, Shelley / Booker, Angela / McDermott, Meghan 2008. Mixing the Digital, Social and Cultural: learning, identity and agency in youth participation. In Buckingham, D. (ed.) *Youth, Identity, and Digital Media*. MacArthur Foundation Series on Digital Learning, Cambridge, MA: MIT Press, 185-206.

Heine, Bernd / Kuteva, Tania 2005. *Language Contact and Grammatical Change*. Cambridge: Cambridge University Press.

Jenkins, Jennifer 2000. *The Phonology of English as an International Language*. Oxford: Oxford University Press.

Kachru, Braj B. 1986. *The Alchemy of English: The Spread, Functions, and Models of Non-native Englishes*. Oxford: Pergamon.

Lantolf, James P. (ed.) 2000. *Sociocultural Theory and Second Language Learning*. Oxford: Oxford University Press.

Lantolf, James P. / Thorne, Steven L. 2006. *Sociocultural Theory and the Genesis of Second Language Development*. Oxford: Oxford University Press.

Lantolf, James P. / Thorne, Steven L. 2007. Sociocultural Theory and Second Language Acquisition. In van Patten, Bill / Williams, Jessica (eds) *Explaining Second Language Acquisition*. Cambridge: Cambridge University Press, 201-224.

Lombardo, Linda 1999. Advertising as Motivated Discourse. In Lombardo, Linda /Haarman, Louann / Morley, John / Taylor, Christopher (eds), *Massed Medias*. Milano: LED, 85-156.

Morgan, Marcyliena 2004. Speech Community. In Duranti, Alessandro (ed.) *A Companion to Linguistic Anthropology*. Malden MA: Blackwell.

Ohta, Amy Snyder 2001. *Second Language Acquisition Process in the Classroom: Learning Japanese*. Mahwah, N.J.: Lawrence Erlbaum.

Pit Corder, Stephen 1981. *Error Analysis and Interlanguage*. Oxford: Oxford University Press.

Prodromou, Luke 2008. *English as a Lingua Franca: A Corpus-based Analysis*. London: Continuum.

Quirk, Randolf 1988. The question of standards in the international use of English. In Lowenberg, Peter H. (ed.) *Language Spread and Language Policy.* Washington, DC: Georgetown University Press.

Seidlhofer, Barbara (ed.) 2003. *Controversies in Applied Linguistics.* Oxford: Oxford University Press.

Seidlhofer, Barbara 2004. Research perspectives on teaching English as a lingua franca. *Annual Review of Applied Linguistics.* 24, 209–239.

Selinker, Larry 1972. Interlanguage. *IRAL, 10,* (3), 209-231.

Vygotsky, Lev S. 1962. *Thought and Language.* New York and London: The M.I.T. Press and John Wiley & Sons, Inc.

Vygotsky, Lev S. 1978. *Mind and Society: The Development of Higher Psychological Processes.* Cambridge, MA: Harvard University Press.

Walqui, Aìda / van Lier, Leo 2010. *Scaffolding. The Academic Success of Adolescent English Language Learners.* San Francisco, CA: WestEd.

Wenger, Etienne 1998. *Communities of Practice: Learning, Meaning, and Identity.* Cambridge: Cambridge University Press.

Websites

Wenger Etienne. <http://www.ewenger.com/theory/>.

VOICE 1.0. <http://www.univie.ac.at/voice/page/what_is_voice>.

ELISA CORINO / CRISTINA ONESTI

Agreement and Disagreement in Newsgroup Interaction[1]

1. Introduction

This chapter aims to investigate the use of agreement and disagreement routines in a peculiar form of communicative interaction: the newsgroup (henceforth: NG). It is based on the NUNC[2] (*NG UseNet Corpus*), a suite of multilingual corpora based on the UseNet NG system, providing the linguistic community with up-to-date written data (2002-2006)[3] of a particular form of textual interaction and language. The advantages of the NUNC-UK[4] corpus are many:

a) the availability of a great quantity of textual data that linguists can rely on;

---

1 The authors are jointly responsible for writing the present chapter. In particular, Elisa Corino wrote sections 3, 4, Cristina Onesti wrote sections 1, 2, 5.

2 The NUNC suite was implemented at Turin University by the research group led by Carla Marello and Manuel Barbera and is freely searchable at <www.corpora.unito.it>.

3 All the following examples are drawn from the NUNC corpora. The 2010 data were also made available at the end of 2011 thanks to the *VALERE* Project (Formal Varieties in Newsgroups of European Languages: Structural Features, Interlinguistic Comparison and Teaching Applications) within which the present study was carried out. The research project is devoted to the study of formal varieties of languages, focusing in particular on Computer Mediated Communication from an interlinguistic point of view.

4 Language specific subcorpora have been created out of the general suite; here we are dealing in particular with the one specifically related to messages by English-speaking users from the United Kingdom. The subcorpora NUNC-UK-Engines, NUNC-UK-Business, NUNC-UK-Photography and NUNC-UK-Cooking have respectively 12,426,186, 7,021,782, 722,818, and 1,322,330 tokens. Nonetheless we had the opportunity to take advantage of further data offline.

b) a contemporary lively language presenting a wide variety of styles and topics;
c) the textual interest for 'quoting' phenomena;
d) the lexicographic and sociological interest in UseNet as a "folk taxonomy" (cf. Barbera/Marello 2010).

The core question though involves, prior to the investigation of the textual and pragmatic aspects of the interactions, the definition of a new textual genre, together with its characteristics, its rules, its boundaries and above all its peculiarities compared to more 'traditional' genres. In particular we will focus on the *quoting* mechanism, which turns out to be the most relevant device for creating a fully cohesive and – most of all – coherent text. After discussing some of the textual features of this particular form of CMC (Computer Mediated Communication), trying to define its properties as a 'text', we will focus on the agreement and disagreement scales and on different ways of arguing in online environments, where Netiquette influences the co-construction of texts through discourse interaction. The corpus analysis could not be based only on queries concerning words in the semantic field of 'agreement', as our interests also concern the use of connectors and linguistic structures. A manual search was carried out on the 2002-2006 data read with the *Agent Fortè* newsreader software.

2. Newsgroup as a genre?

Netspeak is not a single homogeneous *genre* of communication, it is rather a set of 'socio-technical modes' that combine the technological means of communication with the social and cultural practices that have arisen around their use. Such kinds of texts seem to elude the traditional definitions of 'text', which were worked out for traditional *Textsorten.*

Since they first became popular (in the '70s and '80s) NGs have established themselves as a peculiar form of communication within CMC: a virtual community of people sharing the same interests and problems, creating a space for themselves, unrelated to other channels, in order to discuss such interests (see Corino 2007), a sort of public discussion board freely available on the Internet, for messages posted by many users from different locations. UseNet can be compared to a series of specialized journals based on grassroots' participation. They have grown rapidly, gradually varying the subjects and now presenting a remarkable subdivision in themes. More precisely NGs are categorized according to *hierarchies,* i.e. thematically organized argumentative frameworks.[5] They group together users who share common interests and experiences. Therefore, beside general 'just for fun' conversations, we can find texts written by experts discussing very specific issues characterised by the use – to a certain extent – of a peculiar LSP (Language for Specific Purpose): philosophy, politics, baseball, bonsai, classical culture, reflexology, a certain singer or his/her opinion, etc. Contributing to the creation and development of an NG gives users the opportunity to make their own space prominent, as a sort of private publishing company with a right to 'publish' thoughts and discussions.

NGs can be subject to trolling or sporadic flame wars.[6] However, they can be a valuable source of information, friendship and support, bringing people together from different backgrounds. The most striking traits of such a kind of communication depend on the nature of the text itself: an NG is a free on-line forum where people can post messages and contribute to the discussion with their personal opinions, thus building complex discursive and argumentative moves, both in generic texts and in specialised subcorpora.

In an oral conversation we can identify repetitions and redundancy which help speakers to compensate for the lack of a text to refer to. On the contrary, the written nature of an NG makes it possible to

---

5 Hierarchies also define the name of newsgroups. The term *top-level hierarchy* refers to the hierarchy defined by the prefix before the first dot (e.g. the newsgroup UK.TECH.DIGITAL-TV.CRYPT would be in the *uk.* top-level usenet hierarchy).

6 *Trolling* means posting off-topic or inflammatory messages in an online community, in order to provoke readers into an emotional response. *Flaming* is an unfriendly and insulting interaction between Internet users, degenerating into flame wars.

define all the entities and objects which are the focus of the speakers/writers through the use of *quoting* devices. Quotations are the foremost way of contributing to the creation of the text itself and serve as a sort of mnemonic support which helps restore the textual/conversational moves. The original text can be (1) totally reproduced; (2) partially deleted and split into different paragraphs; (3) totally omitted. Depending on the choice of the users, different forms of intertextuality are generated: weak in the case of (1) or strong in the case of (2); whilst as for (3) textual coherence is ensured by the title.

As can be inferred from the sequences of nicknames in any *Agent* screenshot, sparring match conversations as well as group discussions can be clearly identified: conversations can take place between only two participants; however, we can also see the intervention of a third participant starting a new conversation with the first poster or even with both the previous ones. In all these cases the messages of the NG show a constant resumption of the original text of previous messages or parts of them. The quoted sections can be graphically identified by different colours and/or by the automatic insertion of angled brackets (or double, triple angled brackets when quoting an already quoted section, and so on). This process works as a sort of Chinese box mechanism where, within a single thread in a particular 'genre' of NG, we can find self-sufficient microtexts which are formed by the use of quoting together with the comments on the quoted passages. The quoting contributes to assigning a new semantic significance and pragmatic function to the text, depending on the receiver's understanding of the message and intentions, thus his/her quoting choices. Internal coherence is also ensured by the title of the thread and by the fact that all the authors basically stick to the core theme. Topic shifts are generally limited to short digressions.[7]

The focus mechanism changes according to the nature of the discourse, whether it is a *task oriented* dialogue or casual conversation. In the first case we assume all the statements to be relevant and informative, whereas in the second case the linguistic event is the result of the collaboration of the participants negotiating the focus of the discourse. NGs share the features of both cases, integrating task-

7 This contrasts with the situation outlined in Lambiase (2010).

oriented dialogue with casual conversation, although texts are mainly well organized and structured.

In comparison with other CMC genres, posts in NGs are usually longer and more structured. Users' exchanges are complex, with multiple responses often directed at a single initiating message or single messages responding to more than one message. This is a typical feature of asynchronous CMC, and we argue that in this environment the quoting mechanism is the most popular way of achieving successful communication.

3. Style variation and online communication

Studies on language in CMC concentrate in many cases on graphic peculiarities (see for example Gheno 2004) and on Internet users' lexicon, although style variation is often neglected despite some recent approaches (Berruto 2005; Rainer/Stein 2003). There is ample scope for linguistic studies to be carried out starting from this kind of language and textual genre, especially in the field of discourse analysis and argumentation, considering style variation and the connection between the language used, the theme and such a peculiar medium and environment as the NG is.

On the other hand in some linguistic fields there is still little attention paid to CMC as a resource; theoretical linguistic studies still have to investigate how language is used in this particular context of communication. For instance previous studies on (dis)agreement and argumentation, which emerged in very different fields, in particular within law and legal studies, psychotherapy and focus groups, have not considered this kind of 'written conversation' – or 'conversational writing', which can shed light on some of the most recent trends in contemporary use of language, ranging from discourse structure to microsyntactic phenomena.

NGs present a semi-formal context where people are free to express their opinions, but they are also asked to argue and back their claims. People always have to keep to the Netiquette and to the informal rules of the community; flames are frowned upon. These three peculiarities have to be taken into account in our study about agreement and disagreement.

In such an environment where dense interaction takes place and many people are involved, even agreeing and disagreeing strategies seem to follow certain patterns and rules: this study aims at analysing such strategies and the linguistic devices involved. The argumentative moves of English threads are here considered, paying attention to the structures of complex messages (in terms of length, interaction, and argumentative chains). Two features are examined: the logical structure, i.e. the argumentative frame that determines the style and the structure of the discourse (thesis, argument, reinforcement…), and its formal realization in terms of syntactic elements. Among the characteristics which distinguish CMC, such as the extensive use of focalizers and mitigation strategies to express agreement and disagreement, the *quoting* device marks a peculiar feature of the textual genre 'NG', influencing the referential movement of the discourse and, in a broader sense, the construction of the text as a whole.

Last but not least, social rules have to be taken into account as male and female posters can be expected to follow different discursive patterns, just as they do in spoken discourse, both in the way they structure the message and quote passages, and in the way they express (dis)agreement and select the content of their messages. We have not analysed 'the lower profile of women' studied by Pedersen/Macafee (2007) concerning the blogosphere, i.e. "more personal content and orientation towards the social aspects of blogging, as opposed to a male emphasis on information; lesser technical sophistication; and a greater preference for anonymity", NGs being more related to specialized topics than blogs.

In general, though, we can state that agreement and disagreement strategies and discourse development in an interactive environment such as the thread of an NG also display the different features of the so called 'genderlects': women often disagree by cushioning their disagreements with affiliative comments, posting questions rather than making assertions, whereas men use an adversarial style. Nonetheless some patterns of disagreement which are usually considered to be typical of female posters might prove to be more language-specific and culture-related than gender-specific.

4. Different levels of agreement

According to Bazzanella (1996) levels of agreement and disagreement can be mainly divided into three areas:

- total agreement,
- partial agreement,
- total disagreement

in different degrees, as illustrated in Table 1 below:

Total agreement
Sharing
Partial agreement
Repeated term = lexical suggestion
Quotation
Knowledge acquisition
Turn taking strategy
Information used as topic
Answer to a question
Asking for clarification, explanation, motivation
Asking for correction
Amazement
Ironic intention
Partial disagreement
Total disagreement

Table 1. Levels of agreement (Bazzanella 1996).

Although not all the possibilities mentioned by Bazzanella can be easily traced back in NGs, the following examples[8] show the realization

8 Examples are reproduced in their authentic form. Possible mistakes are due to NG users (or to tokenization devices, as regards punctuation). The significant linguistic elements for our analysis are marked in italics.

of many different possibilities. A qualitative evaluation is given, without quantitative purposes. Example 1 underlines total agreement among the users, following a positive statement ('Yes, I agree with all above', accompanied by exclamation marks as another clear indication of a positive attitude towards the interlocutor):

(1) > Soya sauce - different brand / manufacturer tastes not the same , even
> though they are all (light) ' soya sauce '. Hence , unless the recipe
> specifies one particular brand -- normally not specified -- the result is a
> hit and miss. There are many more of mine . Have you got any other
> pat hate ? Bee . --
Hi Bee Yes , *I agree with all the above* !!

Adverbs can state total agreement with the same strength: *I agree with you entirely*, which emphasizes the full understanding and sharing of the expressed idea. Example 2 goes further, stating that *everyone* probably agrees with the opinion mentioned, beyond the subjective opinion, giving a sort of universal and almost objective view about the sticking of rice discussed in a NG devoted to cooking:

(2) > Whether light soya sauce is added depends how dry the mass is . I can't
> recall I ever used ginger , sesame oil , or MSG in this dish . 5) And , of
> course , the meat and veg can be of any sort around at the time . My
> favourites are Char Siu , ham (sold diced) , peas (from frozen packs)
> or ..... just too many to name for contrasting taste and texture .
One thing I think everyone agrees : the tough coat of rice sticking tothe wok or the bottom of the pan is just too dif ...... " Well , whoseturn is it to wash up tonight ?"

In (3) an agreement about the same principles is maintained, although 'different scenarios' are underlined:

(3) […] and enhance their own ego. Only if the client lets them. And if they do, they probably fail in other areas of business management as well
After all , it 's usually men who design products that have more knobs , sliders and flashy lights that , when you come to think of it , have been left in the the position when the product was first bought . As they say " I know you can do it but do you need it ?"
We're talking about different scenarios, but probably agreeing about the same principles

A distinct perspective, probably due to different working situations, does not undermine agreement about general facts of business management. The adversative coordinate clause lowers the strength of 'difference' given in the premise, trying to maintain a general positive mood in the community. The sense of community, even of virtual brotherhood in some cases, is one of most powerful driving forces in such a context, as already remarked in Section 2 (see Corino 2007). Many linguistic strategies aim at defending or supporting others' *face* (as understood in politeness theory; cf. Goffman 1955) and mitigating threatening attitudes. Mitigated verbal forms can occur very frequently (see the common use of *would* or other conditional forms):

(4) It seems that the conclusion is that besan flour is obtained from grinding chickpeas , channa , garbanzo , etc , Any difference in colour / texture that is observed may be due to whether the chickpea is roasted before grinding . I think it is important to get this right . At least half of India rely on gram/dhall for their protein/iron . Together with rice , there is a balance of amino acids (present in meat) and essential for human well being . We should get this right . *I would agree with Ian* , (as usual)

(5) The chillies, garlic and ginger I'm impressed with but the spice front is a bit lacking ;o) snip *I would agree (uh ! I'm agreeing with Si .... ; ?)* much more spice!

Furthermore the role of connectives in agreement and disagreement strategies is crucial. The same verbs as in the previous examples can be immediately rectified by a following adversative conjunction, as in (6-8): *but* and *however* are the most common. Here the writer stresses his/her partial agreement rather than the concepts not agreed on, again avoiding 'face threatening acts' as much as possible (Brown/Levinson 1987 [1978]). It was surprising not to find occurrences of *nonetheless* within the field of adversatives, as it is frequently used in texts with formal style connectives.

(6) > It is polite , and safer , to credit the source of the publication , as you
> do in the recipes following your post , which puts one in the same
> position as a reviewer . Reviews are specifically exempted by
> copyright rules
I agree . *But frankly* I get fed up with the way that every time someone does something positive in this NG , there are always a bunch of whinging twits who find some excuse to criticise .

There 's NO obligation to quote sources , though it's polite , *I agree* . *However* , if you don 't HAVE sources , then you can't .

(7) Eat all you can... *I do agree* , *but* they do prove particularly useful when you are run off your toes and need a quick lunch that is more than a sandwich .

The two parts signalled in (8) by connectives (in italics) can also be more distant in the text, giving the author the opportunity to better argue his/her position, explaining the reasons for agreement and only subsequently the points of disagreement:

(8) "Richard Ashton" wrote:
"Buridan" wrote:
> }You are a time waster. You have been challenged again
> }and again to present rational arguments in support of your
> }atheistic claims, but are evidently unable to do so. Instead
> }you resort to blustering rhetoric, which serves as a smoke
> }screen for you inability to present arguments.
>
> There is not any evidence, credible on examination by the
> scientific method, to postulate the existence of any gods.

I agree that there is no scientific evidence which supports theism (neither atheism). That is because theism is a metaphysical explanation, like materialism, which (by definition) is not empirically verifiable. That does not mean, *however,* that there are not reasons for supposing theism to be true. Some of these I referred to earlier.

In an attempt to re-frame an opinion, the agreement can also be modified by an *if*-clause, stating specific conditions for the agreement itself and clearly defining its boundaries:

(9) > Your overall ignorance is breathtaking . Since there are digital SLRs
> (really , there are !) posts about them are germane to this group .
> Maybe almost all digital cameras have through the lens viewing , but
> the point and shoots do it with a low res , or relatively low res LCD , not
> a prism .
Maybe I 'm becoming troll sensitized , but that , to me , didn 't have the ring of irony to it . If it were , indeed , meant ironically , then I apologize for my rant .
... *I agree* , *if he meant it* . But I took his post as leg pulling .

Talking about agreement and disagreement strategies, some studies have focused on the use of delay devices to introduce disagreement after some apparent agreement (Pomerantz 1975). Other interesting forms of agreement and mitigation expressions could be traced in the following examples:

(10) > by the photoshop 'gang' is that nothing compares . Therefore
> people are driven to theft . [...]
We seem to be agreeing here :-)
Whether the solution is ' free ' software [...], or affordable software is a matter of individual choice . *Accept that I will never agree* with the phrase ' driven to theft '. This is an argument often used to support 'free ' software , but it just seems like twaddle to me . Theft is theft .

The use of emoticons like ":-)" accompanies and reinforces here the mitigating value of the verb *seem*; this is consistent with the main use of emoticons in NG and CMC in general. The locative *here* narrows down the scope of agreement, which is actually limited, and then denied by a peremptory adverb (*never*) as concerns the phrase *driven to theft*.

(11) I checked with out local TS , and they said they would investigate a complaint such as this . 1 And *I 'm quite sure they 'd reach the same conclusion as I have . Now if the ingredients weren 't clearly labelled I'd agree with you , but they are , and I simply can 't see a problem with it .*

Disagreement in (11) is delayed and mitigated by the *if clause* which introduces a potential agreement in a hypothetically different context. The adversative conjunction instead is emphatic about the actual state of things and clearly defines disagreement, though softened by the adverb *simply.* See also the highlighting of *never* in example (12), which is rendered through a graphic device, the asterisks:

(12) > yes , lime pickle is bitter like other citrus fruits (but it does mellow
> on maturing , so may be perceived by some as sour) , reen or unripe
> mango is sour (like pomegranate) . There is the difference.
It may be a difference between different people 's interpretations , but *I would * never * class* a citrus fruit as " bitter ". *I would say* that lemon (juice) is the quintesscence of ' bitter '. [...] well either you or I have an atypical opinion on what is what !! *but I would agree* that people 's perceptions on food are different

Some examples of total agreement show the use of quoting that involves a loss of textual connectives when expressing agreement, as in the case below:

(13) at 08:45 PM, <xxx@easynet.co.uk> said:
> There are a number of questions here which I am sure the others will
> answer better than I can, but why use needle bearings when a plain
> phosphor-bronze bearing would be better, have more surface area
> and would retain libricant better than a needle ?

I agree with Peter – having just replaced bearings in the rear trailing arms of my Fiat Tempra estate (prior to a holiday on Irish roads!) I can vouch for the fact that small diameter bearings don't seem to last. These were taper roller bearings with of course their own inner races, *but* the small angular movement & relatively high loads seem to break the rollers *nevertheless.* I also remember that Mini's had similar trouble - they had a needle roller on the outside & a bronze bush on the inside & on mine at least the needles went first.

These examples underline again how important the thread is: the text of a single post finds its coherence within the entire thread (i.e. in all the postings preceding it). As shown in many examples, textual coherence is also maintained through regular recourse to quoting, a fairly economical device used for constructing argumentation. As Marello (2007) points out, quoting itself determines the use of anaphors and textual deixis, thus making even short messages significant and consistent according to the Gricean laws (Grice 1975).

When talking about 'partial agreement' we note that it is often produced from an agreement which is functional to expressing some level of disagreement. The analysis of connectives shows that this strategy is pursued by using adversative linkers, in particular *however* as example (14) illustrates:

(14) > Unfortunately when tax revenues are raised, by duties on drugs, the
> government effectively becomes a pusher, so any regulations
> introduced will always be ineffective ones. As is the case now.

I agree that regulations would not be 100% effective. They would *however* prevent drugs being marketed in a way that increases their use tremendously – as said, I would expect that *no* advertising of any kind would be permitted

Baym (1996) deals with the difference between male and female participants in online discourse. Where female participants' disagreement included quoting, it was linked to previous discourse, and had pervasive and extensive elaboration. Accounts and justifications emerged with agreements, and not just disagreements, as the notion of preference predicts (Coates 1986).

When analysing NG texts it is not really possible to single out in a reliable way whether the poster hiding behind a nickname is a male or a female writer, although there are some hints which can suggest their gender. It turns out that women are more polite than men, make a more extensive use of mitigation strategies, and tend to quote longer passages in order to make sure their posts are clear. On the other hand, men are more direct and do not care about being offensive. In the following Italian passage, for instance, a male participant is answering a female one (with a very short quoted part) asking for information about a kind of flour. He stresses very precisely (almost pedantically) what he had already mentioned in a previous message:

(15) *Q:* > *Semola*, DXX. *Semola* di grano duro (venduta anche come > *sfarinato di grano duro*).
A: No, stellina; semolino-semolino, quello che si usa per fare le pappate ai microbimbini, che ha la grana simile a qualla dalla farina di mais bramata bergamasca, che viene venduto in pacchettini da 250/500 g o sfuso da sacchi da 50 kg. proprio quello, addizionato di circa un 20 % di quella che viene venduta come semola di granoduro ma che a me sembra proprio *farina-farina* di grano duro. Fo molti errori, moltissimi di questi molti sono voluti, gli altri sono caz....uali, ma in questo caso giustissimo fui.[9]

Moreover, studies report that quoting is more frequent when the participants disagree partially, whereas it is likely to disappear in mes-

---

9 *Q:* > *Semolina*, DXX. *Semolina* of durum wheat (also known as *flour of durum wheat*).
A: *No, little star; middlings-middlings*, the one you use to make babyfood, that is similar to the maize meal from Bergamo, that is sold in 250/500 g packets or by measure 50 kg sack. Exactly that one, with 20% of what is sold as semolina of durum wheat but looks like *flour-flour* of durum wheat. I make a lot of mistakes, some of them are deliberate, others are caz....ual, but this time I was really right.

sages where participants disagree totally (Pistolesi 2004). In the case of disagreement, where we do have quoting instead, any other linguistic device disappears: there are no connectives, adverbs or conditionals as analyzed before. In (16) answers follow one another with unfriendly coldness (see the first message with double angled brackets, then another writer under angled brackets, and a third interlocutor at the end). The argumentative power comes solely from ideas and facts, without any rhetoric. The final question is not rhetorical at all: the user wants to make a precise objection.

(16) >> The british pacifists and appeaser society, also known as the
>> cowards who refused to defend their country, is still alive I see.

> This is a typical response. The fact is Iraq posed no threat to
> Britian or the US. Its now clear that intelligence was inaccurate and
> there is no evidence of WMD. Those where simply excuses. This war
> isn't about defending Britian its about control of the middle east
> which western nations have a vested interest in. Its the same old
> colonisation story you think Britian would learn from its past.

And why in Iraq a war and not in the rest of the world that's waiting
for (re-)colonisation too?

These data show how the quoting mechanism has become pivotal in NGs' interaction, more than in other CMC genres: "Therefore, quoting acquires a more strategic, argumentative role in newsgroups as compared to the email data, where it serves mainly to help the reader contextualize the reply" (Severinson Eklundh 2010).

5. Conclusion

The quoting mechanism alters the textual development of NGs, characterizing very specifically and more than other linguistic devices the textual type of NG threads. Contrary to other cases in the World Wide Web (Storrer 2002; Fiorentino 2005), coherence devices are to be found in the development of more NG messages, with a sort of *a parte subiecti* interpretation, as Conte (1999) has pointed out from a very clear text linguistics point of view. The general sense of the text is created by subjects, interacting with the community beyond the single post.

In such a context agreement and disagreement strategies are not solely based on the traditional devices we can find in oral dialogues or written blogs. Users' interactivity and cooperation occur through recourse to anaphors and several (co)references, showing the same – though 'virtual' – proximity that orality has, according to Koch (1993) (see the dichotomy communication distance vs. immediacy in Koch and Österreicher's model), with some interesting differences between male and female writers. However, the main distinctive feature in conveying (dis)agreement in newsgroup interaction lies in the quoting mechanisms and in their different management.

A further study on diaphasic variation could add additional elements to this first description, since until now CMC has been analysed prevalently for its informal style and language sloppiness, thus neglecting the case of experts talking in a community of highly specific interests with extensively developed argumentative moves. The recently expanding sociolinguistic branch of research that carries out investigations within corpus linguistics (Romaine 2008) shows promising insights which could be further developed by quantitative data. In the future we would like to investigate in greater depth the current situation of NGs with more data, analysing more recent materials and the possible changes due to a modified situation, since other means of communication are currently prevailing on UseNet.

References

Barbera, Manuel / Marello, Carla 2010. Tra scritto-parlato, Umgangssprache e comunicazione in rete: i corpora NUNC. *Studi di grammatica italiana*. XXVII, 2008 = Per Giovanni Nencioni.

Convegno Internazionale di Studi. Pisa/Firenze, 4-5 Maggio 2009, edited by Antoni, Anna / Stefanelli, Stefania, 157-185.

Baym, Nancy 1996. Agreements and disagreements in a computer-mediated discussion. *Research on Language and Social Interaction.* 29/4, 315-345.

Bazzanella, Carla 1996. *Repetition in Dialogue*. Tübingen: Niemeyer.

Berruto, Gaetano 2005. Italiano parlato e comunicazione mediata dal computer. In Hölker, Klaus / Maaß, Christiane (eds) 2005, *Aspetti dell'italiano parlato. Tra lingua nazionale e varietà regionali,* Proceedings of the Congress in Hannover, May 12-13 2003. Münster/Hamburg/London: LIT-Verlag *Romanistische Linguistik* 6, 109-124.

Brown, Penelope / Levinson, Stephen C. 1987 [1978]. *Politeness: Some Universals in Language Usage*. Cambridge: Cambridge University Press.

Coates, Jennifer 1986. *Women, Men and Language*. Harlow, Essex: Longman.

Conte, Maria-Elisabeth 1999. *Condizioni di coerenza. Ricerche di linguistica testuale*. Alessandria: Edizioni dell'Orso (I ed. 1988, Firenze: La Nuova Italia).

Corino, Elisa 2007. NUNC est disputandum. In Barbera, Manuel / Corino, Elisa / Onesti, Cristina *Corpora e linguistica in rete*. Perugia: Guerra, 225-252.

Fiorentino, Giuliana 2005. Così lontano, così vicino: coerenza e coesione testuale nella scrittura in rete. In Korzen, Iørn (ed.) *Lingua, cultura e intercultura: l'italiano e le altre lingue. Proceedings of the 8th SILFI International Congress, Copenaghen, June 2004.* Frederiksberg: Samfundslitteratur.

Gheno, Vera 2004. Prime osservazioni sulla grammatica dei gruppi di discussione telematici di lingua italiana. *Studi di Grammatica Italiana.* XXII, 267-308.

Goffman, Erving 1955. On Face-Work: An Analysis of Ritual Elements in Social Interaction. *Psychiatry: Journal of Interpersonal Relations*. 18/3, 213-231.

Grice, Paul 1975. Logic and conversation. In Cole, Peter / Morgan, Jerry L. (eds) *Syntax and Semantics 3: Speech Acts*. Academic Press: New York, 41-58.

Koch, Peter 1993. Oralité médiale et conceptionnelle dans les cultures écrites. In Pontecorvo, Clotilde / Blanche-Benveniste, Claire (eds) *Proceedings of the Workshop on Orality versus Literacy: Concepts, Methods and Data.* ESF, 225-248.

Lambiase, Jacqueline J. 2010. Hanging by a Thread: Topic Development and Death in an Online Discussion of Breaking News. *Language@Internet.* 7, article 9.

Marello, Carla 2007. Does NGs 'quoting' kill or enhance other types of anaphors? In Korzen, Iørn / Lundquist, Lita (eds) *Comparing Anaphors. Between Sentences, Text and Languages.* Frederiksberg: Samfundslitteratur Press, 145-159.

Pedersen, Sarah / Macafee, Caroline 2007. Gender differences in British blogging. *Journal of Computer-Mediated Communication.* 12/4, article 16, <http://jcmc.indiana.edu/vol12/issue4/pedersen.html>.

Pistolesi, Elena 2004. *Il parlar spedito. L'italiano di chat, e-mail e sms*, Padova: Esedra.

Pomerantz, Anita M. 1975. *A Study of Some Features of Agreements/ Disagreements.* PhD dissertation. Irvine: University of California.

Rainer, Franz / Stein, Achim (eds) 2003. *I nuovi media come strumenti per la ricerca linguistica.* Frankfurt am Main: Peter Lang.

Romaine, Suzanne 2008. Corpus linguistics and sociolinguistics. In Lüdeling, Anke / Kytö, Merja (eds) *Corpus Linguistics. An International Handbook*, vol. I, Berlin: de Gruyter, 96-110.

Severinson Eklundh, K. 2010. To Quote or Not to Quote: Setting the Context for Computer-Mediated Dialogues. *Language@Internet*, 7, article 5.

Storrer, Angelika 2002. Coherence in Text and Hypertext. *Document Design* III, 156-168.

Websites

Corpora UniTo <www.corpora.unito.it>.
NUNC home <http://www.bmanuel.org/projects/ng-HOME.html>.

Exploring the Blogosphere

GIULIANA GARZONE

Where Do Web Genres Come from? The Case of Blogs

1. Introduction

Developments in computer-mediated communication (CMC) and the increasing importance and pervasiveness of web applications are certainly among the factors that have played a most crucial role in determining the proliferation of non-literary discourse genres and their unprecedentedly rapid evolution in the contemporary world. These developments have also posed a challenge to genre studies, and led them to question the suitability of their methodological toolkit for the analysis of web-mediated texts, suggesting the need to adjust them to the analysis and categorization of phenomena presenting peculiarities never dealt with before.

Already in the early 1990s Yates and Orlikowski (1992) discussed the problems associated with the categorization and analysis of web-mediated genres, suggesting that they may derive from the reproduction or adaptation of existing communicative genres, or emerge in totally new forms, thus making important additions to existing genre repertoires and creating new ones (Orlikowski/Yates 1994; cf. also Crowston/Williams 2000).

In an earlier work (Garzone 2007: 16-17) I distinguished between two different kinds of 'genre migration' to the web (as I proposed to call the transfer of pre-existing genres to the web): some genres – e.g. the online glossary, the web encyclopedia, the online newspaper – are preliminarily submitted to a degree of modification and adaptation before being turned into the electronic/hypertextual format to meet the peculiarities of the new medium (Crowston/Williams 2000; Lemke 2002), while others (e.g. the e-book, the academic article in pdf format, the tourist brochure) are transferred to the web as they are, with no apparent preliminary adjustments, although in time the new medium inevitably induces forms of adaptation and change. A third category includes web genres that are native to the web, and therefore specifically tailored to the affordances of the electronic medium: this is the case with emails, homepages, websites, hotlists, newsgroups, and – above all – all web 2.0 applications, from Facebook to Twitter.

Therefore, in order to track the process of genesis or evolution of web genres and identify their antecedents, in the first two cases it is necessary to explore traditional communication formats in order to find pre-existing genres from which electronic/multimodal genres have descended (cf. e.g. Orlikowsi/Yates 1994: 547, and Garzone 2002 on e-commerce), whatever the degree of adaptation to the computer-mediated environment they have subsequently undergone. As regards the third category, i.e. new web-native genres (e.g. emails, blogs, social networks), the problem is different as by definition they have unprecedented peculiarities that set them apart from all traditional genres (Askehave/Ellerup Nielsen 2004; Garzone 2007).

Of course, it is widely recognized that new genres do not emerge from scratch. As Todorov clarifies in his attempt to answer the question 'Where do genres come from?' genres quite simply come from other genres: "a new genre is always the transformation of an earlier one, or of several: by inversion, by displacement, by combination" (Todorov 1976/2000: 197). Although new genres do not derive directly from one or more existing genres, for all of them it is possible to recognize 'antecedents', or 'ancestral genres' (cf. also Herring/Scheidt *et al.* 2005: 144). According to Herring/Scheidt *et al.* (2005), this is also true for web-mediated genres, some of which can be recognised to be "at least partially 'reproduced' in the sense of Crowston and Williams (2000)" (Herring/Scheidt *et al.* 2005: 144). Thus in principle it should be possible to track also the origins of any of the new web-mediated genres by tracing them back to one or more pre-existing genres.

In this chapter the question of web genres' antecedents and of the mechanisms at work in the rise and evolution of web-native genres will be examined by means of a case study – the weblog or blog – a web genre that has evolved rapidly, giving rise to a multiplicity of discursive phenomena (Miller/Shepherd 2009). This genre has been considered particularly relevant not only in consideration of its particular suitability for discussing the question of antecedence, but also because – as Giltrow and Stein (2009: 14) point out – "blogging sparks questions central to genre theories".

The basic framework of this work is in discourse analysis, with special consideration given to genre studies, and within them, to theorizations focusing on web-mediated communication and more specifically on blogs (cf. Section §2 below). The textual component will also be examined carefully, in the conviction that linguistic/discursive clues can provide evidence of phenomena and trends.

The chapter is organized as follows. In the next section (§2) the genre chosen as a case study, the weblog, will be briefly introduced, before going on to deal with the problem of the identification of its antecedents (§2.1) also in light of the relevant literature. This issue will then be discussed looking at two types of organizational blogs, the 'news and blogging' site (§2.2) and the corporate blog (§2.3). This will be followed by discussion and conclusions (§3).

2. Weblogs and their genesis: a case study

Although some early analysts – mostly authors who were bloggers themselves and involved in software development – considered the blog to be uniquely digital and native to the web, descending directly from 'what's new?' or 'cool links' pages and the relevant hotlists (e.g. Blood 2000), there is wide agreement today that, in spite of the novel and unique features inherent in the medium and the technological solutions it is based on, it is still possible "to look for the chromosomal imprint of [its] ancestral genres" (Jamieson 1973), i.e. traditional offline genres. Of course, the sheer size of the phenomenon makes it problematic to identify trends and prevailing features that apply to all blogs indiscriminately, given that as of 31 May 2011 there were over 162 million public blogs in existence.[1] But a good degree of generalization is possible.

1 Cf. *Blog Pulse*. The Nielsen Company. <blogpulse.com> (retrieved on 31 May 2011).

Before going on to discuss the issue of blogs' genre antecedents, it is worth recalling briefly the main characteristics of the weblog as a genre. Its currently accepted definition is based on three main constitutive features: the reverse chronology of its entries, the frequent updating, and the combination of links with personal commentary (cf. e.g. Miller/Shepherd 2004: 4; Herring/Kouper *et al.* 2005: 1). These properties are associated with recurrent peripheral features that make the blog promptly recognizable: blogs present posts, which include a date, a time stamp, and a permalink,[2] usually followed by a repertoire of links; most of them feature comments, which react to or comment on the content of relevant posts (cf. Riboni 2010a: 91ff.). It is the combination of posts, commentary and links that can be considered to be distinctive of the genre, making it possible to realize a peculiar kind of interpersonal interaction and favouring a sense of community among bloggers. It has also been suggested that a further distinctive trait is each blog's participation in "a massively distributed but completely connected conversation covering every imaginable topic of interest" (Marlow 2004; cf. also Blood 2002) engendered by the presence of links to other blogs, although it has been shown that not all blogs contain links to other blogs (Herring/Kouper *et al.* 2005), but often rather to websites, to mainstream media or to social networks (e.g. Facebook) (Myers 2009: 28-37).

As regards the rhetorical action they accomplish, Miller and Shepherd (2004) see blogs as a good example of how an emergent genre, offering a new rhetorical opportunity made possible by technological advances, came to life and became very popular in a surprisingly short span of time as it met the rhetorical needs of society in a specific moment of its cultural and ideological history. The rise of the weblog in the 1990s served the rhetorical exigences of an age characterised by the blurring of the divide between private and public, as epitomised by President Clinton's relationship with the general public and the media, and by the rise of reality television:

2 A permalink, or permanent link, is a URL that points to a blog post and remains unchanged, even after it has passed from the front page to the archives.

> The cultural moment in which the blog appeared is a *kairos* that has shifted the boundary between the public and the private and the relationship between mediated and unmediated experience. (Miller/Shepherd 2004: 4)

This cultural moment – which for some aspects opened a stage in Western culture that is still with us – was characterised by a sort of collective 'mediated voyeurism' (Calvert 2000) and at the same time an aspiration to self-disclosure.

Within this interpretative framework, Miller and Shepherd (2004) focus principally on the personal blog and see it as the response to "a widely shared, recurrent need for cultivation and validation of the self, [...] located at the intersection of the private and public realms." This exigence is fulfilled through the four functions of self-disclosure: self-clarification, social validation, relationship development, and social control (Miller/Shepherd 2004: 9). Self-expression on such a large scale is made possible by the peculiar affordances of the blog format, which has been shown to be in a bridging position between HTML documents (the standard means of communication on the WWW), relatively static and asymmetrical, and interactive text-based Computer Mediated Communication (e.g. newsgroups, chatrooms, etc.): blogs "allow authors to experience social interaction in ways that are otherwise difficult to achieve through web pages, while giving them ownership of, and control over, the communication space that is difficult to achieve in CMC (Herring/Scheidt *et al.* 2005: 162). This is represented graphically in the following figure from Herring/Scheidt *et al.* (2005: 161):

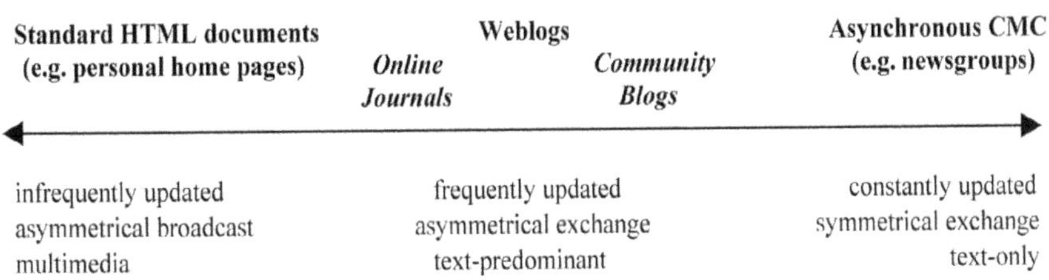

Figure 1. Weblogs on a continuum between standard HTML documents and CMC (Herring/Scheidt *et al.* 2005: 161).

2.1. Blogs' antecedents

Discussing the blog's family tree, Miller and Shepherd (2004) recognize the imprints of a whole range of different ancestors, which they divide into three main groups: first, the ship's log (as also hinted at by the name 'weblog', devised by John Barge in 1997), the commonplace book, the clipping service (or media monitoring service), and the curio collection[3] or the scrapbook – i.e. books similar to the anthology or the Italian *Zibaldone*, containing a compilation of sundry sorts of observations, opinions and facts, integrated with pictures, small objects and other 'treasures', all genres that keep a chronological record of thoughts, facts and objects; second, the journal and the diary, genres used for immediate self-expression (cf. McNeill 2003; Riboni 2010b); third, the pamphlet or broadside, the editorial and the opinion column, in which authors put forth their opinions and thoughts.

Herring/Scheidt *et al.* (2005) also share this personal, existential view of blogs, to the point of complaining that "the importance of blogs as individualistic, intimate forms of self expression" are underestimated (Herring/Scheidt *et al.* 2005: 142, abstract). Similarly to Miller/Shepherd's (2004), this view is part of a general trend to emphasise the individual, existential dimension of weblogs, giving prevalent attention to personal, journal- and diary-type blogs. However, in the same essay Herring/Scheidt *et al.* (2005: 159-160) also highlight the diversification of blogs as a function of the topics they deal with, identifying for each different type the influence of a specific antecedent, e.g. travelogues in the case of travel blogs, and researchers' written project journals in the case of k-logs,[4] etc., as well as the influence of other digital genres.

3 It can be inferred that the 'curio' Miller and Shepherd mention amongst the antecedents of the weblog is an album collecting paper samples, picture cards and cuttings.

4 'K-log' is an acronym for *k*nowledge b*log*; it is a type of blog generally used as an internal/Intranet blog, not accessible to the general public, which serves as a knowledge management system. The term 'klog' is also used to describe a blog that is technical-content oriented. Cf. <http://www.webopedia.com/TERM/K/klog.html> (retrieved 31/05/2011).

The emphasis laid in research on the individual, existential dimension of blogs is warranted on the one hand by the history of their evolution and the characteristics that have prevailed in the course of their history, which Miller/Shepherd (2009: 266-268) divide into three main stages: before 1999, when blogs were used mainly by 'web savvy individuals' and mainly consisted in links to sites of interest on the web and comments on such sites; a second phase, when the number of users grew and the focus shifted to personal expression; and a third phase, showing a strong continuity with the previous one, but an intermingling with social networks. Still today personal blogs prevail in the blogosphere in purely numerical terms, as is confirmed by a look at the chart below, retrieved from BlogPulse Live.[5]

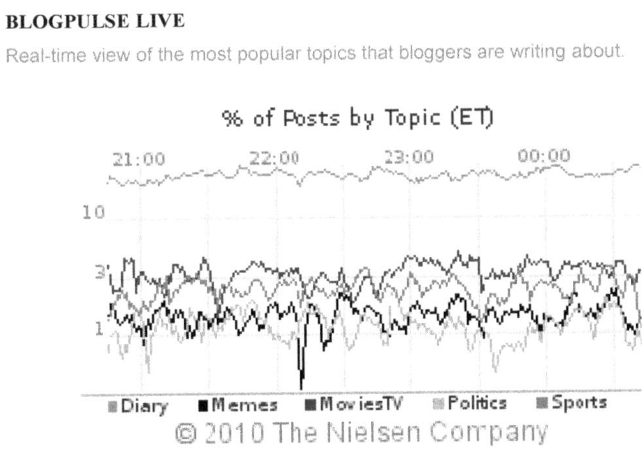

Figure 2. Most popular topics, 31 May 2011.

Figure 2 shows that the number of blogs in the diary/personal journal format, represented by the line at the top of the graph, is still by far the highest, while the numbers of blogs dealing respectively with cultural issues, TV and movies, politics and sports are more similar in inci-

5 BlogPulse Live, powered by The Nielsen Company, "provides a real-time view of the most popular topics that bloggers are writing about, updated by the minute and presented as a percentage of all blog posts." Cf <http://www. blogpulse.com/> (retrieved on 31 May 2011).

dence, with the relevant lines in the lower part of the chart occasionally overlapping or intertwining.

However, in terms of importance and influence the picture is quite different, because in the last few years other types of blogs have increased their relative prominence, with many of them evolving into more complex discursive forms resulting from the contributions of several bloggers, and many others being started on the websites of institutions, organisations, e.g. companies, public departments, research bodies, newspapers or magazines, etc. Besides, as Herring/Scheidt *et al.* (2005) predicted a few years ago, the blog software and the blog format are increasingly being utilized in a diversified range of communicative uses well beyond the original borders of the blogosphere, in politics, journalism, business, popularization, domain-specific information (e.g. about health, cars, entertainment, gossip, etc.) on account of its peculiar properties that make it an attractive sociotechnical format for a wide range of purposes.

Many of the Technorati Top 100 blogs were originally founded by one person for the purpose of sharing his/her opinion with the world but, as their success grew, blogs became ever more complex in their structure and organization, involving teams of writers to produce posts. This is especially the case with the types of blogs that Miller/Shepherd (2009) call 'public affairs blog', including opinion blogs, political and news blogs, which according to them are highly influential in national politics. In some cases the news blogs have grown into full-fledged online newspapers, and some of them are hosted on the website of newspapers – like the *Los Angeles Times* or the *New York Times* – or other newsmakers, like *CNN*, having been started, or acquired, by a news company.

A meaningful example is that of no. 1 in the Technorati Top 100 blogs, *The Huffington Post,* which is an interesting case of radical transformation of the genre, so that it provides the opportunity to ask whether it still has the distinctive traits of the blog. Founded in 2005 by Arianna Huffington and others as a commentary outlet and an alternative to news websites, in time it has grown into a 'news and blogging' site, and now offers coverage of politics, media, business, entertainment, living, style, the green movement, world news, and comedy, with several local versions for different US cities and an international version. The posts authored by thousands of bloggers, including politicians, celebrities, academics and policy experts are integrated by original news content and featuring columns by editor-in-chief Huffington and a core group of contributors. *The Huffington Post* was acquired by AOL on February 7, 2011, and has since then increased its popularity. In the next section this site will be analysed as a case study, looking at the organization of its content and at the characteristics of the texts posted on it, with a view to identifying antecedents and other possible influences, and verifying whether it can still be classified as a blog.

2.2. The news and blogging site

The Huffington Post describes itself as '*The* Internet Newspaper', and this representation is confirmed by the layout of the homepage, organized in the typical style of online newspapers, with a block title at the top accompanied by a large photograph taking up the whole width of the page, and below it displaying the posts in three columns. Text distribution is similar to that of online newspapers, too, following the typical inverted pyramid pattern: the long scrolling homepage features links for as many as a hundred posts or articles, mostly consisting of a headline and a picture, located in an inner layer of the hypertext. The page also hosts various menus with hyperlinks to internal and external features and sources, as is typical of blogs.

It is more than evident that the materials published on the site are the result of hybridization of the original personal diary-like format of the blog with news discourse, thus confirming the site's self description as a newspaper. What impact the hybridization between these two orders of generic resources has had on texts can be inferred from a close analysis of the materials, which seem to be relatively diversified in discursive terms.

Let me start by looking at the *Featured Blog Posts*, presented in the lower part of the left column of every page of the site, which constitute the blog proper. The posts are introduced by means of a title accompanied by a paragraph of text (the lead) and exhibit the typical peripheral features of the blog, i.e. the blogger's picture and synthetic biographical details (e.g. "Larry Flynt, Publisher of Hustler magazine and free speech advocate"; "Jim Wallis, Christian leader for social change"). These posts, sometimes also authored by public figures or celebrities, typically tend to adopt a discursive approach that is similar to that of editorials, a journalistic genre whose purpose is widely recognised to be that of "countribut[ing] to the moulding of public opinion on current affairs" (Westin 2002: 101). Editorials are typically centered on a problem-solution pattern (Hoey 1986), with an initial part defining the problem, followed by evaluative statements and a conclusive section putting forth recommendations or expectations about future developments (van Dijk 1992: 244; Vestergaard 2000; Westin 2002). This is the pattern followed for instance in a post by Sen. Tom Udall,[6] entitled 'Libya Policy is a Slippery Slope', which starts with a short sequence introducing the problem:

(1) Over 60 days ago, President Obama notified the Congress that he intended to use military force in Libya. In the face of an imminent humanitarian catastrophe, it was the right thing to do and I supported it.
Along with the international community, the United States was clear – the targeting of civilians by Muammar Gaddafi would not be tolerated.
But this was intended to be a limited military operation. After 60 days, I fear that we are adrift. We are without direction and we are in danger of fighting a slowly expanding war, now pushing for regime change.

This opening is apparently informative, but the facts exposed are common knowledge and here they are only reported to prepare the ground for what follows. This corresponds exactly to the first move in editorials, which has been described as 'Providing background information' (cf. Ansari/Babaii 2004). Subsequently the author's view of the problem is put forth, thus defining the issue to be discussed:

(2) We have been down this path before and it's a slippery one.
If our military is to be involved in an escalation, then Congress must exercise its Constitutional authority and approve or disapprove of the President's proposal.

---

6 Udall introduces himself as "New Mexico's 17th United States Senator". The text was posted on 12/06/2011.

This stance is then supported by means of arguments and facts (the military escalation is dangerous, the military actions that are being conducted go well beyond those authorised by congress, we do not know enough about the rebels to be wanting to arm them, etc.). The exposition is mainly in the first person plural with *we* obviously referring to the American nation.

The text meets many of the requisites that in the literature have been recognized as typical of editorials, in terms of structure – as illustrated above (Vestergaard 2000) – and in terms of linguistic realization, with recurrent features such as the consistent use of modals like *must* and *should* to indicate obligation and *would* for hypotheses ("Congress *must* exercise its Constitutional authority…"; "But the military goal *should* be defined …"; "Flooding the region with small arms *would* be a major mistake…"), recourse to rhetorical questions to dramatize the argument ("Who's to say those arms won't end up in the hands of some very unsavory and dangerous individuals?"), and use of editorial *we* – in this case referring to the American nation – to fictionally encode consensus and invoke solidarity (Fowler 1991: 208-221; Hawes/Thomas 1996).

But at various points there is special emphasis on the personal dimension, and the symmetrical activation of an interpersonal component:

(3) Believe me, I deplore Muammar Gaddafi. I support a democratic transition and his departure from power. But the military goal should be defined and limited as a matter of policy. It should not include regime change. That is a dangerous escalation.

This casts Udall's views in a strongly personal and interpersonal perspective, which emerges again immediately before his conclusions:

(4) Those are my opinions and there are others in the Congress that must be heard. I have been proud to serve New Mexico as a U.S. representative and then U.S. senator for more than a decade. I have served in the Congress during two wars and I have seen the impacts on our military, on their families and on our national deficit.

Were it not for the maxim of Relation (cf. Grice 1989: 27), this statement would be at odds with the contents of the text. In actual fact, it is aimed at striking a typically ethical note, as Udall talks about his loyalty to the country to add force to his stance, thus setting the ground for his final peroration:

(5) Before the United States escalates its involvement in another overseas conflict, this body must weigh in. It's our constitutional duty to our country and to the people.

This is in line with van Dijk's (1992) view that the closes of editorials typically feature either expectations about future developments, or normative opinions, viz., recommendations, about what actors should do, or not do in connection with the problem being debated (cf. van Dijk 1992: 244).

In contrast, the news content that integrates the blog is not different from that presented on daily newspapers' websites. On the homepage only the headline and a picture are given for every piece of news, and the story – accessible by means of a hyperlink – is usually organised according to the conventional criteria applied in the press. In most cases the only possible difference with news articles is the presence of a picture of the author (usually an in-house reporter) in the abstract of the story, which of course does not appear when the story comes from a news agency (e.g. Reuters).

An example is that of a story entitled 'Yoplait Pulls Ad Said To Promote Eating Disorders' by reporter Laura Stampler (posted on 15/06/2011) which starts with the typical news report lead, giving the gist:

(6) Yoplait has *agreed to pull* a yogurt commercial that the *National Eating Disorders Association* (NEDA) believes could trigger dangerous behavior in those suffering from eating disorders.

The story continues with the description of the commercial, which is also available in video format, and with an interview with NEDA's president, followed by a narration of how the commercial came to the attention of NEDA, and by other interviews. The final paragraph, as is typical of news articles (van Dijk 1988), provides background information about NEDA's activity:

(7) NEDA has fought what they describe as 'David versus Goliath' battles against numerous companies whose ads – often unintentionally – encourage an unhealthy relationship with food.

This analysis confirms that this story has the typical organization of the news article, as information is not put forth sequentially and logically, but rather in 'instalments' (cf. van Dijk 1988) and organized in a typically non-linear pattern that has been conceptualized in the literature as an 'inverted pyramid' (Bell 1991: 168-169; Conley 2002) or as an 'orbital structure' (cf. Iedema/Feez/White 1994; White 1998; cf. also White 2000: 388-389).

What, however, is typical of the blog is the huge number of comments elicited by every piece. In this case, only a few hours after being posted the story about Yoplait had aroused over 1,500 comments.

2.2.1. Comments

Comments are often considered to be a constitutive component of the weblog as a genre, although there are some that do not have them. Organized as they are in a reverse sequence, from which independent strands of discussion occasionally depart, in discursive terms they certainly seem to owe a lot to chatrooms, newsrooms and forums, i.e. to interactive social media. In actual fact, comments do have some traits of the mixed oral/written modality characterising CMC in general (cf. among others Baron 2000; Crystal 2001/2006): heavy use of first and second person pronouns, present tense, contractions, a relatively low level of formality, the occasional use of capitalisation for 'shouting', as in the following comment, which also contains a rude expression ("for Christ's sake"):

(8) Samantha Marie[7]
19 minutes ago (12:13 AM)
This is completely ridiculous! How the HECK is comparing something like making a healthy decision (celery, exercise, yogurt) opposed to making an unhealthy eating choice (eating cake) PROMOTING AN EATING DISORDER?!? For Christ's sake, in the country of obesity running rapid, at

---

7 In quotations from posts underlined text indicates hyperlinks.

least they're promoting making a healthier eating decision! AND thinking of exercise! NOT depriving yourself of food!

But other traits of social media, such as the use of initialisms and vulgar or violent expressions are rarer, if not totally absent, in the blogs analysed here. And in many cases texts are well-organized and the range of lexical choices wide, with some syntactic complexity in longer comments. See another example from the comments to the Yoplait story:

(9) jtjrfan8
06:42 PM on 6/15/2011
Well if they pull this ad then I want all make up ads pulled! I have very deep facial scars on my cheeks from a fire I escaped from 40 years ago as a child! It hurts me to see smooth and airbrushed skin on the tv! I also want all pain medication ads removed because with half of my joints being replacements-it offends me that some one on tv tells me if I take their over the counter pill (as opposed to my prescribed pain management ones) I won't hurt! [...] You know the rules; we cannot favor one group over another! So the commercials which offend me will be off the air as of when?!

Apart from its argumentative complexity and syntactic intricateness, this excerpt shows very well that it is in comments that the original spirit of the weblog is still preserved. Whatever the topic, people send in comments to express their own opinions and feelings, often taking a personal, existential approach.

The other peculiar element that emerges strongly in comments is the interpersonal component, with real debates among commenters. The following sequence is reproduced here in the original order and layout in which the comments appear in the blog, with ganerd's comment (12) that has triggered MichaelRCooke's reaction (10), and Lisa Shields's inserted comment (11) posted as a reply to the MichaelRCooke's:

(10) MichaelRCooke
4 hours ago (4:04 AM)
Actually no, for you metaphor to be on point, it would be like selling an alcoholic energy drink by using the interior monologue of an alcoholic who has to have a drink, but fears not being alert enough for the business meeting –

I and a younger guy showing him the way by buying the alcoholic energy drink! Gee – problem solved! (not really, not at all).

(11) HUFFPOST SUPER USER
Lisa Shields
Poet & Advocate For Special Needs Children
0 minute ago (8:25 AM)
You're a touch behind the times.
Ever seen a television ad for scotch? Gin? Rum? ANY hard liquor?
They stopped running those ages ago.
Wine and beer, yes...hard stuff, no.

(12) ganerd
6 hours ago (2:30 AM)
LOL Ironic, since Yoplait has SO MANY chemicals, it is a highly processed food. I would never eat it if I was given a free years supply. If you care about your health, you will eat an organic yogurt like Stonyfield farms like I do. Check out a few of Yoplait's chemicals and the damage they can do at Dr. Mercola's MD website <mercola.com>.

This sequence illustrates very well the mixed modality of comments. In particular comment (12) has capitalization for shouting, a punctuation totally based on oral language, and the use of 'LOL', the best known of all Internet initialisms, which stands for 'laughing out loud' (Crystal 2001/2006: 34).

The examination of comments to various posts shows that language use and register in comments is also sensitive to topic, the more serious and engaging the topic, the less informal the comment. Here are a few comments to Sen. Udall's post analysed above:

(13) aearthling59
01:27 PM on 6/12/2011
Get out of NATO, SEATO and the UN. Sink or swim.

(14) 01:26 PM on 6/12/2011
WASHINGTON (AFP) – The US military has flown more than 800 sorties over Libya since handing control of the air campaign's operations to NATO, the Defense Department said Tuesday (Apr 11)
Did congress declare war yet? No. This is against our constitution and an impeachable offense to take America to war without a declaration from congress.

(15) MaryLouWho
08:34 AM on 6/12/2011
Just get out of Libya now.

Dealing with a political topic, the personal opinion component prevails over the existential one and, in spite of the important interpersonal orientation, the commenters' linguistic choices are more on the written side; in this respect, the quotation from AFP in (14) is quite meaningful.

In the following section, attention will shift to the other case study presented here, that of the corporate blog.

2.3. The corporate blog

In the corporate world, weblogs have had an extraordinary success story in the last few years. They are used mainly in two different ways.[8] Within the business enterprise, as Kolari *et al.* (2007) point out, blogs are radically changing internal communication, empowering employees to voice their opinions publicly as a form of corporate democracy, to share expertise and initiate discussions. Thus blogs improve collaboration and enable internal business intelligence. But blogs are also a powerful tool in external corporate communication, from public relations to marketing, being useful for customer fidelization, customer service, product promotion, business related news, as they enable the company to keep an individualized dialogue going with each single stakeholder, and in particular with actual and potential customers.

As an example we shall look at Kodak's blogs. The company manages three: *Grow Your Biz*, addressed to corporate customers, *Plugged In*, introduced as "a place for ideas and knowledge sharing from the people of Kodak about technology", and *A Thousand Words*, introduced as "a place for stories from the people of Kodak." For all three blogs, in the initial presentation there is reference to Kodak's passion for customers and for images and technology ("We love what

8 Of course, within the two broadly defined areas of use of corporate blogs, there is a wide diversification of uses. For a discussion cf. Puschmann 2009: 54ff.

we do, and we want to share our stories/our experience about imaging and its power to influence our world.").

The latter blog is particularly interesting, being presented as a conversation place where personal stories are shared. This approach exploits the distinctive individualistic, existential properties of the genre mainly to promote the sale of Kodak products. Contributors to the blog are all Kodak employees, and they generally write pieces where they seem to share their personal thoughts with people, while at the same time they promote Kodak's products. A typical entry in this blog is the piece posted by Worldwide Marketing, Retail Systems Solution Director, Rowan Lawson, on June 17, 2011 entitled 'It's Graduation Time – And Father's Day Too. So here's what this Dad wants…', in which he sets out very personal thoughts about life and paternity at a time when his second daughter is leaving school, and asks questions about his performance as a father ("Did I spend enough time with her? […] Was I there for her when her first relationship broke down?" etc.), and then concludes by recommending a photo album as the best gift for Father's Day:

(16) So this year, this Father's Day has far more meaning than usual. Normally I'd be happy with a power tool, a pair of socks, or a part for my motorcycle. But this time it's different. I've realized that I've been cruising along, only to find that the speedometer of life has been showing 1% of the *real* speed we're going […]. But this Father's Day I want something simple that I can view in my own time. I want to look back on the last 17 years of my family with kids.

On the occasion of Father's Day, a very good opportunity for Kodak to increase its sales in the U.S., many other similar posts figure on the site. Thus, thanks to the inherent properties of the blog as a genre, product promotion is clad as a personal expression of one's own existential feelings.

While the drafting style of the posts may help make this commercial intent less evident, an analysis of Kodak's website can reveal the company's real use of the blog. The post 'Kodak Cameras Featured in New Kings of Leon Music Video' (posted on 26/05/2011) by Jenny Cisney, Kodak's Chief Blogger, provides an interesting example. This is how it starts:

(17) Check it out! We are giving Kings of Leon fans an exclusive sneak peek at a behind the scenes making of the music video of the award winning band's hit summer song, 'Back Down South'. Kodak armed 80 fans featured in the video with the latest Kodak PlaySport Video Camera and Kodak EasyShare Touch Digital Camera. […]

The post goes on to describe the two new cameras and how they are used, and ends with a personal concluding line.

(18) With this new song to enjoy…we are ready for summer!

An interesting finding is to discover that on the very same day in Kodak's Newsroom a press release appeared entitled 'Kodak Cameras Featured in New Kings of Leon Music Video', with exactly the same text, and the same photograph, with the addition of boilerplate features ('About Kodak').

This provides a good clue as to the antecedents of external corporate blogs, indicating that blogs are perceived only as one among many tools in the company's external communication effort and are managed in the same way as press releases and other corporate discourse types, relying on previous knowledge about genres that are superficially perceived as similar, although in line of principle they are not.

In this context it is not surprising that these posts attract hardly any comments at all. The 'Kings of Leon' post prompted no comments, while Rowan's had eight, all of them very personal, for instance:

(19) What a wonderfully reflective piece. Your 5 questions caused me to do a personal inventory. I think I will pass this post along to my family. Perhaps they will get the answer to the perennial question 'Dad, what do you want for Father's Day' (Answer kindly provided here!).

Of course, it is legitimate to doubt the authenticity of this kind of comments, which probably are as artificial as the posts to which they respond.

3. Discussion and conclusion

The analysis of two representative examples suggests that the blog has evolved well beyond its original function of blurring the divide between private and public suggested by Miller and Shepherd (2004: 9), as organizations have appropriated the format and use it for their own purposes. More specifically, in 'blogging and news sites' the blog genre is hybridized with various forms of news discourse, and in particular the editorial and the news story. Furthermore, in this type of sites the blog proper incorporates editorial news content, which is published alongside 'real' posts in 'post-like' format, with some of the peripheral features characterizing blogs, so that at first sight the inattentive reader may not realize the difference.

As regards external corporate blogs, the analysis has shown that companies often use them for marketing and PR purposes, as they are an ideal instrument to combine world-wide reach and a direct personal contact with each single individual within the global audience. In this case the personal/existential dimension distinctive of blogs is often obtained by featuring texts where the bloggers – mostly corporate employees – deal with issues having to do with their own private life, or at least so they pretend. But it is more than obvious that the copy comes from the marketing or PR team, as it maintains the textual features of press releases and other corporate genres (e.g. advertising, advertorials, company brochures), using employees as figureheads.

The elements gathered in the two case studies confirm that blogs have developed into a variety of forms, in many respects profoundly different from each other, so much as to cast doubts on the generic integrity of this communicative format. In their 2009 paper, Miller and Shepherd go so far as to argue that "the blog […] is a technology, a medium, a constellation of affordances – and not a genre" (Miller/ Shepherd 2009: 283). According to them, in its heyday back in the late 1990s blogging technology was perceived to fit a particular exigence of the period, and the perfect fit between the genre and the medium, the social action and its instrumentality led people to mistake one for the other. But, they continue, in time "the suite of affordances called blogging was discovered to fit other exigences in different ways"

(Miller/Shepherd 2009: 283), new different blogs emerged and the coincidence between the genre and the medium dissolved.

This interpretation seems far-fetched in various respects. First, it obviously rests on a definition of medium which is so broad that in a way it neutralizes the heuristic validity of the notion. Let us consider some definitions of medium given in modern communication theory, e.g. Mc Luhan's (1964/1994: 4) definition as "any extension of ourselves", or more broadly, "any new technology", or Kittler's (1997) scientific definition of medium as a form of data storage and transmission, or Meyerowitz's (1985: 15) definition as a communication system that creates new social environments and is thus active in bringing about social change. All these definitions are located at a higher level of generalization (e.g. the electronic medium, the computer medium, the hypertextual medium, etc.) than that of the genre, and imply that the medium in which each genre is set will have an impact on it and contribute to determining its distinctive features. But this is far from arguing that there is coincidence between medium, with its technological affordances, and genre.

Secondly, the blog is not supported by one technology, but by various technologies which are not qualitatively different from those used in other forms of CMC, websites and social networks, which means that similar technological affordances can give rise to different generic realizations.

Therefore, it is certainly more reasonable to see the blog as a 'macrogenre', as proposed by Bhatia (2005: 34-37) for the letter genre,[9] whose realizations have bi- or polyrhematic denominations specifying the sub-genre to which they belong (e.g. business letter, personal letter, etc.). This somehow helps solve the problem posed by Giltrow and Stein (2009: 9-10) when they observe that with Internet genres one often has to deal with discursive realizations for which it is difficult to decide whether they count as a full-fledged genre or a genrelet.

9 Bhatia (2005: 34) describes the 'letter' as a macrogenre, while he considers the 'business letter' as a 'genre colony'. It is to be noted that among scholars there is hardly any agreement as to how to refer to groupings of a number of (sub)genres that share – at least partially, if not exclusively – the purpose each of them serves.

Thus it is proposed here to consider the blog as a macrogenre, comprising a whole range of genres with (as it were) 'hyphenated' denominations (public affairs blog, corporate blog, political blog, etc.). Obviously, all genres constituting the macrogenre have some distinctive features in common that have survived the rapid change of the last few years, and share at least part of their communicative purpose: all sites that call themselves 'blogs' have the superficial and peripheral formal traits typical of blogs which make them promptly recognisable as such and, in terms of content, strive to maintain at least partially a personal/existential component, even at the cost of adding it artificially.

In light of the foregoing, it can be stated that today the weblog has evolved into a (macro)genre which is used in various contexts of situation for purposes that are different from its original ones. However, alongside these new purposes also a distinctive component of its original one has survived which tends to be constantly present, sharing one's personal opinion/experience with addressees, accompanied by a characteristic textual organization suitable to encode it. In terms of generic antecedents, the blog certainly derives from the diary/personal journal genre, but in each single domain it has hybridized with, or has been colonized by, the discourse genres customarily used in that domain, often to the point of risking being overwhelmed by them.

If these observations are brought to bear on the question of web genre antecedents, it can tentatively be concluded that web genres emerge when the availability of new affordances makes it possible to satisfy social and rhetorical exigences that previously were not adequately met, or were not met at all, by means of new communication formats. But when such new communication formats are in place and consolidate into well-defined and recognisable genres, it is inevitable that they start to be used for new purposes. This is first of all due to a sort of attrition with its users' previous generic knowledge, which causes "older practices from related off-line genres [to] carry over into web genres", as predicted by Herring/Scheidt *et al.* (2005). In other cases, it results from the deliberate decision to exploit the potential of the new genre for a different purpose originally pursued by means of other genres, thus deliberately colonizing it. This has happened for instance with emails, which started out as plain and mostly short messages with evident conversational traits (e.g. Gimenez 2000), and have evolved into a macrogenre hosting discursive practices typical of other traditional or computer-mediated genres, because users transfer their knowledge of other genres into them, e.g. using them as formal letters (a practice that is increasing in the age of certified email), or because an organizational decision has been made to use them e.g. for newsletters (Bait 2007), or for quotations or orders.

If these premises are accepted, then genre antecedents have to be sought not only in those genres to which present-day web genres can be traced back, but also in those that contributed along the way to turning them into their current formats and uses through hybridization.

Of course, the repertoire of web genres is so immense that generalizations are difficult to make. A lot of further research into web genres will be necessary before a satisfactory reply to the question 'Where do they come from?' is found. But conceptualizations and efforts at pattern identification can have a precious heuristic value, being useful to systematize existing knowledge and shed light on hitherto unperceived aspects.

References

Ansari, Hasan / Babaii, Esmat 2004. The Generic Integrity of Newspaper Editorials: A Systemic Functional Perspective. *Asian EFL Journal* 6/1, 1-58.

Askehave, Inger / Ellerup Nielsen, Anne 2004. *Webmediated Genres – A Challenge to Traditional Genre Theory. Working Paper nr. 6.* Aarhus: Center for Virksomhedskommunication.

Bait, Miriam 2007. Tackling the Electronic (R)Evolution: Tutorials and Instruction Websites for E-mail Newsletter Writers. In Garzone/Poncini/Catenaccio (eds), 73-90.

Baron, Naomi S. 2000. *Alphabet to Email: How Written English Evolved and Where it's Heading.* London: Routledge.

Bell, Allan 1991. *The Language of News Media.* Oxford; Blackwell.

Bhatia, Vijay K. 2005. Interdiscursivity in Business Letters. In Gillaerts, Paul / Gotti, Maurizio (eds) *Genre Variation in Business Letters*. Bern: Peter Lang, 31-54.

Blood, Rebecca 2000. Weblogs: A History and Perspective, on line at <http://www.rebeccablood.net/essays/weblog_history.html>.

Blood, Rebecca 2002. Weblogs: A History and Perspective. In *We've Got Blog. How Weblogs are Changing our Culture*. Cambridge, MA. Perseus, 7-16.

Calvert, Clay 2000. *Voyeur Nation: Media, Privacy, and Peering in Modern Culture*. Boulder, Colorado: Westview Press.

Conley, David ^{2}2002. *The Daily Miracle: An Introduction to Journalism*. Melbourne: Oxford University Press.

Crowston, Kevin / Williams, Marie 2000. Reproduced and Emergent Genres of Communication on the World Wide Web. *The Information Society.* 16/3, 201-216.

Crystal, David 2001/2006. *Language and the Internet*. Cambridge: Cambridge University Press.

Fowler, Roger 1991. *Language in the News. Discourse and Ideology in the Press*. London: Routledge.

Garzone, Giuliana 2002. Describing E-commerce Communication. Which Models and Categories for Text Analysis?. In Evangelisti, Paola / Ventola, Eija (eds). *TEXTUS* (*English in Academic and Professional Settings*). 14/2, 279-296.

Garzone, Giuliana 2007. Genres, Multimodality and the World-Wide Web: Theoretical Issues. In Garzone/Catenaccio/Poncini (eds), 15-30.

Garzone, Giuliana / Catenaccio, Paola / Poncini, Gina (eds) 2007. *Multimodality in Corporate Communication. Web Genres and Discursive Identity*. Milano: FrancoAngeli.

Giltrow, Janet / Stein, Dieter 2009. Genres in the Internet. Innovation, Evolution and Genre Theory. In Giltrow, Janet / Stein, Dieter (eds) *Genres in the Internet: Issues on the Theory of Genres*. Amsterdam: John Benjamins, 1-25.

Gimenez, Julio 2000. Business E-mail Communication: Some Emerging Tendencies in Register. *English for Specific Purposes* 19, 237-251.

Grice, Paul 1989. Logic and Conversation. In *Studies in the Way of Words*. Cambridge, Mass.: Harvard University Press, 21-40.

Hawes, Thomas / Thomas, Sarah 1996. Rhetorical Use of Theme in Newspaper Editorials. *World Englishes* 15/2, 159-170.

Herring, Susan C. / Kouper, Inna / Paolillo, John C. / Scheidt, Lois Ann / Tyworth, Michael / Welsch, Peter / Wright, Elijah / Yu, Ning 2005. Conversations in the Blogosphere: An Analysis 'From the Bottom Up'. *Proceedings of the 38th Hawaii International Conference on System Sciences.* IEEE.

Herring, Susan C. / Scheidt, Lois Ann / Wright, Elijah / Bonus. Sabrina 2005. Weblogs as a bridging genre. *Information Technology & People* 18/2, 142-171.

Hoey, Michael 1986. Overlapping Patterns of Discourse Organization and their Implications for Clause Relational Analysis of Problem-Solution Texts. In Cooper, Charles / Greenbaum, Sidney (eds) *Studying Writing: Linguistic Approaches*. Beverly Hills, CA: Sage, 187-214.

Iedema, Rick / Feez, Susan / White, Peter R.R. 1994. *Media Literacy*. Sydney: Metropolitan East Disadvantaged Schools Program.

Jamieson, Kathleen M. H. 1973. Generic Constraints and the Rhetorical Situation. *Philosophy and Rhetoric* 6/3, 162-170.

Kittler, Friedrich (1997). *Gramophone, Film, Typewriter*. Stanford: Stanford University Press.

Kolari, Pranam / Finin, Tim / Lyons, Kelly / Yesha, Yelena / Yesha, Yaacov / Perelgut, Stephen / Hawkins, Jen 2007. On the Structure, Properties and Utility of Internal Corporate Blogs. *Proceedings of the International Conference on Weblogs and Social Media* (ICWSM 2007). Boulder, Col. Available at <http://www.icwsm.org/papers/2--Kolari-Finin-Lyons-Yesha-Yesha-Perelgut-Hawkins.pdf>.

Lemke, Jay L. 2002. Travels in Hypermodality. *Visual Communication* 1/3, 299-325.

Marlow, Cameron 2004. Audience, Structure and Authority in the Weblog Community. Paper presented at the 54th Annual Conference of the International Communications Association, New Orleans. Available at <http://cameronmarlow.com/media/ marlow-2004-audience.pdf> (12/10/2011).

McNeill, Laurie 2003. Teaching an Old Genre New Tricks: The Diary on the Internet. *Biography: An Interdisciplinary Quarterly* 26, 24-48.

Meyerowitz, Joshua 1985 *No Sense of Place: The Impact of Electronic Media on Social Behavior*. Oxford: Oxford University Press.

Mc Luhan, Marshall 1964/1994. *Understanding Media. The Extension of Man.* Cambridge MA: MIT Press.

Miller, Carolyn R. / Shepherd, Dawn 2004. Blogging as Social Action: A Genre Analysis of the Weblog. In Gurak, Laura *et al.* (eds) *Into the Blogosphere: Rhetoric, Community, and Culture of Weblogs*. <http://blog.lib.umn.edu/blogosphere/blogs_as_virtual.html>, retrieved 04/08/2010.

Miller, Carolyn R. / Shepherd, Dawn 2009. Questions for Genre Theory from the Blogosphere. In Giltrow, Janet / Stein, Dieter (eds) *Genres in the Internet: Issues on the Theory of Genres*. Amsterdam: John Benjamins, 263-290.

Myers, Greg 2009. *The Discourse of Blogs and Wikis*. London: Continuum.

Orlikowski, Wanda J. /Yates, JoAnne 1994. Genre Repertoire: The Structuring of Communicative Practices in Organizations. *Administrative Sciences Quarterly* 33, 541-574.

Puschmann, Cornelius 2009. Lies at Wal-Mart. Style and the Subversion of Genre in the *Life at Wal-Mart* blog. In Giltrow, Janet / Stein, Dieter (eds) *Genres in the Internet: Issues on the Theory of Genres*. Amsterdam: John Benjamins, 49-84.

Riboni, Giorgia 2010a. *Comunicazione e interazione discorsiva in Rete: Il caso dei blog politici*. Doctoral dissertation. Dottorato in Anglistica. Università degli Studi di Milano.

Riboni, Giorgia 2010b. The Life of Lives: From Paper To Web. In *Culture, Annali del Dipartimento di Lingue e Culture Contemporanee della Facoltà di Scienze Politiche dell'Università degli Studi di Milano.* Milano: Montedit, 275-289.

Todorov, Tzvetan 1976/2000. The Origin of Genres. In Duff, David (ed.) *Modern Genre Theory*. Longman: Harlow (English transl. Catherine Porter, *Genres in Discourse*, 1990; orig. L'origine des genres, 1976).

van Dijk, Teun A. 1988. *News as Discourse*. Hillsdale NJ: Erlbaum.

van Dijk, Teun A. 1992. Racism and Argumentation: 'Race Riot' Rhetoric in Tabloid Editorials. In van Eemeren, Frans H. *et al.* (eds), *Argumentation Illuminated.* Dordrecht: Foris, 242-259.

Vestergaard, Torben 2000. From Genre to Sentence: The Leading Article and its Linguistic Realisation. In Ungerer, Friedrich (ed.) *English Media Texts. Past and Present.* Amsterdam: John Benjamins, 151-176.

Yates, JoAnne / Orlikowski, Wanda 1992. Genres of Organizational Communication: A Structurational Approach to Studying Communication and Media. *Academy of Management Review* 17/2, 299-326.

Westin, Ingrid 2002. *Language Change in English Newspaper Editorials.* Amsterdam: Rodopi.

White, Peter R.R. 1998. *Telling Media Tales: the News Story as Rhetoric.* PhD Dissertation. University of Sidney: Department of Linguistics.

White, Peter R.R. 2000. Media Objectivity and the Rhetoric of News Story Structure. In Ventola, Eija (ed.) *Discourse and Community. Doing Functional Linguistics.* Tübingen: Gunter Narr, 379-397.

SANDRA CAMPAGNA

Antagonizing the Editor: Speech-styles Variation in *The Economist* Reader Comments*

1. Introduction

In Spring 2010 *The Economist* published an editorial in the May 15th issue followed by two related articles on the controversial topic of the Islamic veil in Europe. The articles specifically referred to the (then) recent legislative bills proposed by both the Belgian and French governments in favour of banning the burqa and the niqab in public places in Belgium and France.[1] Not only did the drastic ban raise a heated debate in public opinion at large inside and outside Europe, as expected, but it also affected, on a smaller scale, *The Economist* community. On that occasion 63 readers expressed their viewpoint responding to the editor's strong criticism of the widespread (largely European) consensus on the anti-Islamic veil measures. Furthermore, the debate was amplified by the high degree of readers' interactions, a common trait in the context of online periodicals conducive to Web 2.0 communicative practices such as the ones featuring the weblog as a genre, for example. Thus, although the space allowed to *The Economist* readers is not exactly a 'spontaneous' blog, but rather a controlled forum, the open access to commenters' exchanges justifies the claim advanced in this chapter that the total of reader comments on a

* This study was published with a grant from MIUR PRIN Project n. 2007JCY9Y9 'Tension and Change in English domain-specific Genres' coordinated by Maurizio Gotti; University of Turin project: 'Genre Migration: Intertextuality and Interdiscursivity across Media'.

1 The bill in France was approved by the National Assembly in July 2010 and by the Senate in September 2010. In Belgium the bill only passed the lower house of Parliament in April 2010.

specific editorial (like the ones related to the burqa controversy) can be considered a hybrid form of blogging activity.

This study focuses on how the readers interact with the editor and with each other in this heated debate; what distinctive linguistic/stylistic features characterize the language of commenters in this specific context; to what extent the readers' linguistic traits are representative of the weblog as a genre; and what repercussions (if any) the editor's rhetorical argumentative structure (and language) have on the readers' 'free' communicative styles.

The outline of the chapter is as follows: Section 2 gives an overview of the organizational structure of *The Economist* Leaders and the related Reader Comments by providing information on how topicality is indexed and the average rate of readers' responses to leaders. Section 3 draws on recent studies on Internet genres and focuses on the weblog as a genre; the recurrent features that characterize its generic integrity; how the weblog has evolved/is evolving; and what distinctive traits qualify participatory journalism. Section 4 positions the aims and assumptions of the present study within a theoretical framework. It also situates participatory journalism within *The Economist* context by focusing on the engagement strategies deployed by the editor and the typical argumentative rhetorical structure of editorials. Section 5 provides the linguistic analysis of a sample of comments (constituting a blog) in *The Economist* Leader Reader Comments section (specifically referring to the leader "Banning the Burqa",[2] available online at: <http://www.economist.com/node/16108394>).

In Section 6 the data are contrasted with the language and rhetorical structure of the original leader to investigate whether *The Economist* style has a bearing on the readers' communicative styles or, conversely, whether the language of commenters in this specific environment displays autonomous discursive traits. This is a crucial point in identifying new developments in the blog genre marked by context-specific features.

---

2 Although *The Economist* leader 'Banning the Burqa' only triggered an average rate of bloggers' responses (as reported in Section 2), the blogs attached to this leader have been chosen as a case study because of the topicality of the Islamic veil issue in Europe.

2. *The Economist* leader reader comments: preliminary observations

The Economist leaders[3] display a fairly repetitive organizational structure. First, the number of editorials included in each issue tends to be fixed;[4] second, although all the leaders are important, prominence is given to the main leader (always the first). This is foregrounded by the magazine cover: for example in the May 1st issue the cover indexes the Greek crisis and the controversial role played by Germany as topical by portraying Angela Merkel juxtaposed to the Greek Acropolis. The same image is attached to the main leader: 'Acropolis Now'. Third, there is a tendency in the readers to respond to topicality: usually the prominent editorials correspond to the highest rate of readers' responses, which indicates the editor's indirect strategy to raise a special interest for topical issues. The ten magazine issues under scrutiny (which respectively cover three issues in March 2010, four issues in April and three issues in May) show a high correspondence between indexed topicality and the highest rate of readers' responses (an average of 90%).

Another trait which is worth mentioning is cross-cultural variation in topicality orientation. At times a different cover (foregrounding a different leader) is placed in an edition targeted for a specific country[5] to indicate 'local' newsworthiness. For example, in the March 20th issue, whilst the European, UK and North American editions orient readership by foregrounding the American health reform indexed by the corresponding cover, the Asia-Pacific edition displays a different cover directing the readers towards Thailand's political affairs. Similarly, the March 27th issue and the three May issues in the UK edition

---

3 The quantitative data in this section are based on the observation of a sample of ten issues of the online version of *The Economist* print edition published over two months (March 13th – May 15th, 2010). These issues contain a total number of 57 leaders.

4 In the sample analyzed the number of leaders is 5/6 per issue. Only the March 20th issue contains eight leaders.

5 *The Economist* provides four editions respectively targeted for Europe, the UK, North America and Asia Pacific.

orient the British reader towards the topical theme of the newly-formed coalition in the British government, whilst prominence is given to different subjects in the other three editions.[6] An extended example of cross-cultural variation is found in the May 8th issue where 'local' topicality is signalled through three different covers. However, the incidence of readers' responses to editorials does not always follow the editor's orientations, as shown in Table 1.

Table 1 contrasts the top rated comments in the March 13th – May15th 2010 issues with the leaders foregrounded by the editor. The editorials indexed by the magazine covers are highlighted in italics.

Date	*The Economist* leaders	Top rated comments in the Reader Comments Section
March 13th	*Europe's engine*	630
	Facing up to history	278
March 20th	*Spin, science and climate change*	371
	The battle for Thailand (Asia-Pacific edition)	132
March 27th	*Now what?*	129
	Out of the ruins (UK edition)	39
April 3rd	*Hope at last*	72
April 10th	*Britain's choice*	50
	Sleepwalking towards disaster	121
April 17th	*The new masters of management*	114
	Three years to save the euro (Europe edition)	72
April 24th	*Earthly powers*	60
	Greedy until proven guilty	77
May 1st	*Acropolis now*	154
	Who should govern Britain? (UK edition)	191
May 8th	*Deep trouble*	46
	But can he govern? (UK edition)	16
	Coming to a city near you? (Europe edition)	75
May 15th	*Britain's accidental revolution*	62
	No going back	232
	The wrecking of Venezuela	122

Table 1. Top rated comments related to editorials in *The Economist* March 13th – May15th, 2010 issues in relation to leaders foregrounded by the editor.

6 Despite the change in focus indicated by different covers, the order in which the leaders appear remains unchanged in all editions. Thus the main leader always coincides with the first.

The data displayed in Table 1 demonstrate that the rate of readers' responses, generally speaking, tends to comply with the editor's expectations, but in some cases the orientation highlighted by the editor does not trigger the highest rate of readers' responses. In the April 3rd issue, for example, the top rated comments concern the leader 'Sleepwalking towards disaster', on the Japanese economy, not the prominent leaders, 'Hope at last', on the American health reform system, and 'Britain's choice', on the upcoming election in Britain. A more striking example of readers' 'freedom' of choice appears in the May 15th issue where the main leader on the newly-formed political coalition in Britain ('Britain's accidental revolution') obtained 62 responses (an average number), whilst the top rated comments are associated with two leaders respectively dealing with the future of Europe after the Greek crisis ('No going back') and Chàvez's government in Venezuela ('The wrecking of Venezuela'). The data also show that in some cases the expected global reach suggested by the main leader falls short of the editor's expectations. The leader: 'Facing up to history' in the March 13th issue, for example, accounts for a very high rate of comments even though the topic (Turkey and the Armenian diaspora) concerns a specific 'local' community. On the same note the foregrounded main leader contained in the May 8th issue, 'Deep trouble', on the oil spill in the Gulf of Mexico, triggered only 46 responses (below average), despite its global topicality.

These preliminary observations suggest that the interaction between editor and commenters in The *Economist* environment is mixed: on the one hand, the readers are invited to join the debate even if their choices are definitely dictated by the editor's suggestions; on the other hand, readers make their voices heard also by expressing their preference for unsolicited topics.

3. Internet genres

3.1. The Weblog: a genre, a medium?

In the last decade Internet genres have attracted the attention of researchers and practitioners interested in studying how the Internet setting has been shaping and (re)defining genre. Despite differences in approaching the subject, a common preoccupation underpins this fairly recent scholarly debate: the concept of 'genre migration' and its realizations in the Internet domain. The basic questions which characterize this shared concern are the following ones: What happens when a genre migrates to the Web? Will it retain the same properties that characterized it in the previously represented medium (for instance, the print format)? Does the Web environment generate new native genres or, conversely, does the Web environment host pre-existing genres which nevertheless have had to adapt to the new medium? What are the linguistic/rhetorical features of these emerging/recuperated genres? The fluid and ephemeral Internet terrain does not favour clear-cut answers and often the question of newly-generated/recuperated genres remains unresolved as stated in Giltrow/Stein (2009: 1-2): "The question of genre, then, is really an old one, but the advent of new media has highlighted the issue with new full force. Seemingly, there are new genres on the Internet, but in some cases it is a matter of contention whether the genre is new, or an old one in new medial garb". The blurred divide between emerging and recuperated genres had been anticipated over a decade earlier in Crowston and Williams' (1997) paper on the potential capabilities of Web pages. Crowston and Williams (1997: 3) claim that since genre depends on social acceptance, "it is impossible to define the exact point at which a new genre emerges from the old one".

Another matter of contention defines the Internet/genre issue. This concerns the methodological approach to genre revisited by the new medium. The main question is simply whether a given genre native to the Internet or recuperated by the Internet domain (like the blog, or the chat for example) can be defined on the basis of its recurrent linguistic/rhetorical features or whether formal, textual classifications do not seem to hold with Internet genres due to their close dependence on the flexible Web medium. Also the methodological question (the *text* versus *context* debate) is an old question in genre analysis which, according to Giltrow and Stein, has been recently foregrounded due to the advent of new media. The increasing preoccupation with the medium (and its versatility as the main cause of genre proliferation) has rather discouraged textual/formal approaches to genre (more common in the early days of the Internet) while favouring instead "a revision of the whole set of analytical tools so that the 'media' element is tightly incorporated into the notion of genre, thus determining its overall readjustment as a system" (Garzone 2007: 19).

An invitation to discard *form,* or rather *formalism* as specified in Devitt (2009), and foreground *function* in genre analysis is at the core of Miller's (1984) seminal paper on the concept of genre as 'rhetorical practice'. Miller criticizes taxonomies based on grouping genre items by criteria of similarities and advocates a different approach to genre firmly rooted in societal (evolving) practices. Miller (1984: 163) claims that:

> The understanding of rhetorical genre that I am advocating is based in rhetorical practice, in the conventions of discourse that a society establishes as ways of 'acting together'. It does not lend itself to taxonomy, for genres change, evolve, and decay; the number of genres current in any society is indeterminate and depends upon the complexity and diversity of the society.

Miller's early intuition particularly suits the current Internet scenario marked by constant change (also causing the disappearance of a specific genre besides the emergence of new ones)[7] and a continuous proliferation of fragmented new/remediated genres (like the blog, for example) whose unsettled features can hardly be encompassed in fixed taxonomies.

It follows that in this new rhetorical framework a 'bottom-up' approach to genre seems more efficient than an abstractive 'top-down'

---

7 As Giltrow and Stein (2009: 4) clarify with respect to the 'social action' approach: "Rhetorical theories find these actions in recurring socio-historical situations, rather than abstract categories. Accordingly, from the rhetorical view of the historical contingency of genre, genres expire when the situations expire".

take especially when applied to Internet genres which present a low level of generality and a high degree of specificity (Giltrow/Stein 2009). Inside the newly-formed constellation of Internet genres the Weblog in particular has been a focal object of scholarly attention not only among researchers (McNeill 2005; Devitt 2009; Miller/Shepherd 2009; Myers 2010; to name just a few), but also among practitioners as testified by the wide production of studies (available online) provided by professional bloggers, like Rebecca Blood (definitely an icon in the field), who have contributed to understanding what blogs are, how they can be defined and how they have evolved (Blood 2000, 2003).

Blogs are particularly indicative of the wide debate which characterizes Internet genres and their evolution in that they clearly exemplify the transition from a prevailing interest in identifying recurrent formal/stylistic features which characterize and preserve the blog's 'generic integrity'[8] typical of earlier studies, to a more recent concern for studying what blogs do when placed in newly-generated specific contexts. This shift in emphasis from what a blog *is* to what a blog *does* is a key issue not only framing this specific genre at a micro level, but also reflecting on Internet genres in general and on how these genres constantly change shape (and *do* different things) under the pressure of rapid technological developments as already evidenced in Miller's (1984) paper. This later preoccupation with what the Internet medium allows to *do* and on how newly-constructed *affordances*[9] (new technological possibilities) constantly generate new genre varieties places the issue of the Internet and genre in quite a different perspective: the medium is impinging on genre(s) to such an extent that

---

8 The notion of 'generic integrity' relies on the assumption that genres are recognizable and fairly standardized because they present a 'fixed' set of formal/functional features which define their specificity (Bhatia 2004). In the case of the Internet medium, which is constantly evolving, it is more difficult to identify a genre on the basis of its 'fixed' characteristics.

9 As explained in Garzone (2007: 19): " The term 'affordance' was first coined by the perceptual psychologist James Jerome Gibson in 1966 and further explored in his 1979 book *The Ecological Approach to Visual Perception* (Hillsdale NY. Lawrence Erlbaum). It was adapted to Human-Machine Interaction and later to Human-Computer Interaction by Donald Norman in 1988 in *The Design of Everyday Things* (Jackson, TN, Basic Books)".

centrality on the medium (and its constantly evolving affordances) has become a focal point in the recent relevant literature. For example, the take on blogs advanced in Miller/Shepherd (2009) is particularly innovative with regard to Internet genres: the authors (2009: 283) claim that the blog is: "a technology, a medium, a constellation of affordances – and not a genre".

The weblog (now commonly called 'blog') emerged in the late nineties. The original weblogs were link-driven sites created only by web professionals willing to 'filter' information and share the selected web texts with each other. As noted in Miller/Shepherd (2009: 266) in their early phase these 'filter' blogs shared the following features: "they were chronologically organized, contained links to sites of interest on the web, and provided commentary on the links". Blogs then started evolving and proliferating fast as the Internet affordances grew. The second phase of blogging, as acknowledged in Miller/Shepherd (2009), marked a transition between *filter* blogs and *personal* blogs. The personal dimension of blogging, engaged by younger users, became more prominent; personal blogs were often defined as diaries in digital form (McNeill 2005) because, like in the diary genre, they had dated entries and foregrounded personal narration and commentary. This emphasis on the self dimension of blogging, a phenomenon which should not be underestimated according to Herring *et al.* (2005), has been aptly described by Blood (2000) as follows:

> Free-style blogs are nothing less than an outbreak of self-expression. Each is evidence of a staggering shift from an age of carefully controlled information provided by sanctioned authorities (and artists), to an unprecedented opportunity for individual expression on a worldwide scale.

As specified in Miller/Shepherd (2009) the advent of social networking (in platforms like Friendster, MySpace and Facebook) marked the third phase of blogging currently characterized by what is now commonly referred to as the Web 2.0 platform.[10] Thanks to increasingly

10 This phase corresponds to the third era of Globalization according to Thomas L. Friedman. In his often-cited book, *The World is Flat* (2005), he describes the progression of Globalization 1.0 to Globalization 3.0. We are presently experiencing Globalization 3.0, an era characterized by individuals globaliz-

developing interactive technologies web users can upload (rather than simply download) material on the Internet and share their uploaded content (in the form of linked texts, photos, videos, comments, personal experiences) with other web users. These user-generated-content facilities endorsing blog proliferation also in public contexts (thus moving away from the personal dimension) have re-shaped the genre issue in the Internet domain. Is it still possible to define blogs in formal ways? Are they online diaries? Is the blog's 'generic integrity' still definable on the basis of the following linguistic/rhetorical features generally recognized in earlier studies (mostly based on personal blogs) and summarized in Devitt (2009: 35): dated entries; reverse chronological order; regular updates; hyperlinking; brevity; use of the present tense; self-referentiality; anonymity?

As Myers (2010: 3) suggests: "if blogs are becoming important in political, social and economic life, we need to know how they work, just as we need to know about political speeches, journalism and advertisements. The persuaded have to know what the persuaders are doing".

3.2. Participatory journalism

A wide area representative of how blogging has evolved beyond the personal sphere is what Miller/Shepherd (2009) define as 'public affairs blogs', an umbrella term which incorporates categories falling into the news/politics context, broadly speaking. Most categories relate to the concept of 'participatory journalism', another umbrella term also currently labelled 'citizen journalism' or 'Internet journalism'. Lemann (2006) defines 'citizen journalism' as follows: "The category that inspires the most soaring rhetoric about supplanting traditional news organizations is 'citizen journalism', meaning sites that publish contributions of people who don't have jobs with news organizations but are performing a similar function".

---

ing thanks to the advent of social networks and uploading facilities which allow the dissemination of user-generated content on the Web.

Internet journalism involves active participation in the following settings: 'micro-news' (ordinary citizens reporting on local affairs); 'eyewitness testimony' (ordinary citizens reporting on major world events like 9/11); 'republishing and commentary' (ordinary citizens republishing information originally published elsewhere/commenting on published material); and 'editorial filtering' (ordinary citizens evaluating published articles). A further broad category relates more specifically to the sphere of political opinion. This sphere is characterized by the active participation of bloggers as co-constructors of political consensus often strongly impacting on political campaigns.[11]

A common denominator which characterizes the categories mentioned above is the bloggers' freedom in expressing and sharing individual opinions on news events and political matters often departing from mainstream culture. This has been viewed as a positive contribution to professional journalism (Mitchell 2003) and as an indicative trait of the informal, communicative styles in bloggers contrasting with the 'lecturing' rhetoric typical of American journalism (Gillmor 2003).

However, the perspective of mainstream journalism on bloggers' contributions is not equally positive. Regan (2003) observes that traditional journalists often criticize blogs for being amateurish both in form and content. And, interestingly, participatory journalism is not always evaluated in positive terms even by professional bloggers. As stated in Blood (2003) news reporting by bloggers often lacks originality and cannot replace professional journalism.

4. Participatory journalism in *The Economist* context

As evidenced in Section 3 the complex debate which characterizes Internet genres and their evolution is particularly intricate in the case of the weblog, a genre (a medium?) whose generic integrity is jeopardized by the proliferation of blogs emerging from the original personal blog-

11 The categories included in the 'public affairs blogs' have been adapted from Miller/ Shepherd (2009).

as-a-diary template and taking new forms when located in more structured environments which trigger (and filter?) personal opinions in public affairs as is the case with blogging in some forms of participatory journalism. When blogging is placed in a highly conventionalized context, such as *The Economist*, a preoccupation with *form* (not *formalism* as specified in Devitt 2009) comes to the fore again.

I share Devitt's (2009) concern that a sharp distinction between content and form in genre is not tenable. Form and content are intrinsically linked and as both are determined by context, form (as a determinant of generic integrity) cannot be discarded but needs to be integrated with function: the formal traits which recur in certain types of blogs reveal (and are determined by) the action of blogging in a specific context. In short what blogs *are* depends on (and is constrained by) what blogs *do*. Thus, the formal components of blogs can serve as a reference point to determine genre variation in a specific setting. In this light I hypothesize that there is a close correlation (and mutual dependence) between the Reader Comments blog genre and *The Economist* context; and that the latter acts as a 'proto-genre' and is bound to impinge on the readers' 'free' language. The formal features recognized in the literature as markers of the blog's generic integrity (already listed in Section 3) form a reference point in determining genre variation.

The following key issues will be addressed in the case study presented in Section 5: to what extent the 'proto-genre' conditions the readers' speech-styles in this specific context and which formal features determine the generic integrity of *The Economist* Reader Comments blog genre.

4.2. *The Economist* community online

The Economist community online can be broadly defined as an 'epistemic community' (Riley 2002), a term which draws on previous definitions of 'communities',[12] and focuses particularly on the aspect of

12 In particular on Wenger's (1998) concept of 'communities of practice' characterized by the following converging properties: "mutual engagement; a joint enterprise; a shared repertoire" (Wenger 1998: 73).

sharing common knowledge as a key factor. According to Riley (2002: 47), knowledge informs the individual's *cultural or epistemic competence*, defined as "the sum total of the beliefs, information and skills which one needs to share and apply in the society and the situations in which one finds oneself". In the case of *The Economist* community online, common knowledge and shared expertise qualify the participants both in terms of *what they need to know* in order to access the community successfully and in terms of *what they need to do* (in this case they need to be competent with Internet skills).

The Economist community online is, generally speaking, a three-layer aggregation (formed by the editor, the readers, and the commenters). This authoritative magazine can be defined as 'semi-specialized' in that it addresses economic, cultural, scientific, and technical issues targeted for a rather sophisticated audience.

It is a well-known fact that *The Economist* has a special style characterized by: combining different registers (formal/informal; serious/humorous); alternating arguments and counterarguments in editorials interspersed by forms of direct address to engage the reader; employing metaphorical constructs (a typical trait in specialized discourse as observed in Gotti 2005); using sophisticated language and creating obscure titles rich in extra-textual references and puns. As noted by Bowker (2008), using headlines ambiguously and jokingly is a recurrent feature which characterizes the complex style of *The Economist*. By *bantering* (interacting jokingly) from the very beginning with its sophisticated reader *The Economist* involves the interlocutor in an exclusive 'dialogic language game'. This playful interaction is also a strategy to lighten the complexity of the article which follows the headline (Bowker 2008: 181-182):

> A major strategy used by *The Economist* is play with dislocation of genre, juxtaposing different discourse worlds, and drawing on 'deep' socio-cultural referents and allusions. The reader needs to access a considerable amount of background knowledge in order to solve 'The Dialogic Language Game', and success depends on the culturally specific repertoire possessed by any one individual.

The commenters' 'epistemic competence' (what they need to do in order to 'join *The Economist* club') is specified by the editor in the Reader

Comments section attached to each online article. The readers interested in contributing have a 15-day period to do so. After the deadline the reader comments related to a specific article are 'frozen' and readers can enjoy the commenters' contributions (but they are not allowed to join the debate) and are free to sort the comments as they wish: oldest first/newest first/readers' most recommended.

5. Bantering with the editor: the commenters' speech-styles

5.1. The blog generic integrity: conformity/dissonance

The sample of 63 comments in response to *The Economist* editorial 'Banning the burqa' is analyzed here to identify the extent to which the 'protogenre', in this case the editorial brand, influences the commenters' speech-styles and what formal features determine the generic integrity of the blog in this specific context. The latter are examined in the light of the distinctive traits listed by Devitt (2009: 35) to characterize the generic integrity of the blog: dated entries; reverse chronological order; regular updates; hyperlinking; brevity; prevailing use of the present tense; self-referentiality indicated by recourse to the first person singular; anonymity marked by the extensive use of pseudonyms.

The selection of comments examined reflects only in part these recurrent traits. First, with regard to the temporal dimension, whilst the commenters comply with the convention of organizing their uploads in dated entries, they do not need to conform to reverse chronology which in this context becomes an option allowed by the editor, as highlighted in Section 4 above. Moreover, it has also been noted that the readers cannot post their comments to *The Economist* articles beyond the deadline, thus regular updates are limited to the two weeks' time imposed by the editorial mandate. Second, the generic feature of hyperlinking is virtually non-existent here except for references to the editorial in question. However, it is worth mentioning that the sample reveals an extensive use of extra-textual references taking various forms such as quotations, proverbs or personal narrative digressions.

The sample reports elements reflecting the personal dimension, thus supporting the view expressed in Herring *et al.* (2005) that the personal component in blogs is an important feature also outside the personal blog domain. However, the recurrent linguistic traits which define the personal dimension in the sample conform only in part to the markers of self-referentiality evidenced in the literature. The argumentative discourse (mostly structured impersonally) which characterizes the debate between the readers and the editor, is often interspersed with narrative digressions. The prevailing use of the present tense in these narratives marks conformity to the generic integrity of the blog but this is counterbalanced by a frequent recourse to the past tense mode, which, again, indicates only partial conformity. By contrast, it is worth noting a trend in the commenters to remain anonymous also in this specific context in that all the bloggers use pseudonyms.

Finally, as far as the brevity of comments is concerned, data indicate that 37 blogs out of 63 can be considered relatively brief (maximum length: ten lines). Comment length tends to vary: from a minimum number of about 20 words to a maximum number of 500 words.

5.2. The commenters' speech-styles: distinctive traits

The cases of conformity and dissonance highlighted in the selected sample reveal that the recurrent communicative modes used by the contributors in the specific context are conducive to the following discursive macro-areas: argumentative discourse; personal narrative; and extra-textual dimension. A recurrent rhetorical strategy used by the commenters mainly to contrast with the editor's stance consists in dismantling the editor's argumentative line by developing their own counter-arguments. This process is realized by quoting *The Economist* argumentative statement and reporting salient citations in the attempt to construct a 'counter-editorial', as evidenced in the following examples:

(1) dihydrogenoxide wrote:
May 13th 2010 3:52 GMT
"People cannot, in most countries, walk the streets naked." Indeed, however in most countries a man may walk the streets topless, whereas a woman may not. Imagine the French farce of a woman walking down a Parisian street with a Burqa/Niqab head-dress, but otherwise topless. *Could she be subsequently arrested on two counts of indecent exposure by the Authorities?*

(2) Tom Silo wrote:
May 13th 2010 3:59 GMT
You are wrong, wrong, wrong and don't even seem to get why. The Economist: "indeed, arguably less because the covering was for reasons of faith, not fashion." *TS: Why would religion have any more stature than anything else? Does it deserve some sort of place in our society where it cannot be criticized or discussed? What leads to this idiotic "conclusion" of yours?*

(3) Res Publica wrote:
May 13th 2010 5:02 GMT
--"As Voltaire might have said, "I disapprove of your dress, but I will defend to the death your right to wear it." True. And Montesquieu added: "My freedom ends where the others' freedom begins." *In other words, respect and you will be respected. Is that that another "ethnocentric" western value??? Doesn't it apply the same way in Islam???*

(4) n4b3e1 wrote:
May 13th 2010 7:08 GMT
"Islam is a cults belief that blatantly profess that those who follow are destined to paradise and the rest who don't are condemned to hell." On the contrary, Islam teaches peace, and that the final decision of who enters what rests with God himself. (Unless, of course, your talking about violent individuals who claim to be Muslims--much like a combative footballer claiming to play the game of Football.) Thanks for your pearl of wisdom nonetheless.

(5) J. Kemp wrote:
May 15th 2010 3:31 GMT
"...the burqa makes it harder for police to identify security risks." Yes. Let's have Muslim women who are acting as suicide bombers wear clothing similar to all other French women. *Then they will be quite easy to identify! Is there a word in French for something which* **is** *stupid beyond description?* Maybe "burqa ban" will become the phrase which takes on that meaning in France. Unbelievable.

(6) Super Weh wrote:
May 16th 2010 6:46 GMT
"And although many, and probably most, Muslim women wear the headscarf out of choice" *Has the economist researched this?* This is a pretty bold statement to make without any backing up by research.

A frequent trait which characterizes the commenters' counter-editorials consists in addressing the editor directly through provocative questions as shown in the examples above highlighted in italics. The polemic tone also characterizes their interactive exchanges where provocation, often accompanied by informal linguistic traits, either involves the whole community of *The Economist* online (formed by the commenters, the readers, and the editor), as shown in example (7), or it is used to criticize the editor directly as shown in example (8).

(7) biaix wrote:
May 18th 2010 10:06 GMT
May I ask if are we all nuts? Have you all seen the woman on the picture? Is that normal? Healthy?

(8) Boloyna wrote:
May 18th 2010 8:04 GMT
This article is the epitome of ultra-liberal garbage. The Economist gets worse little by little every year.

As previously noted the sample displays elements enhancing the personal dimension, a typical trait in blogging especially used in the earlier stage of its development. However, in this context the personal dimension is often embedded in the contributors' argumentative line as a digression which in fact is used to corroborate claims by reporting direct individual experience, as shown in the examples below:

(9) mdoaleh wrote:
May 13th 2010 6:03 GMT
The issue of banning the Burqa is at most hypocritical. My daughter one time decided freely to don the Hijab during her first year in university. The second year, she took it off and donned a diamond nose stud. In her third year she volunteered for a one year English teacher in China. During all this she was in the basketball and volley ball teams of her college wearing the normal shorts and jerseys. She is now a well rounded 26 year old government economist. I

am sure that If I or anybody else intervened with her decisions while she was seeking a religious understanding, it would have been counter-productive.

(10) Lafiel wrote:
May 13th 2010 4:57 GMT
Last I checked Europe was not built upon immigration, and therefore has no precedent to cater towards them. Why should it be wrong to defend your home culture against a very foreign one?

(11) Bill_T wrote:
May 14th 2010 9:04 GMT
In the 1940's and 50's my teacher's in Ireland wore something very like a burqa. Their lower faces were visible, their chins barely so. Their habits were a Christian invention from about the 14th/15th century and followed by certain catholic religious orders into the 20th. The burqa is a more extreme form of the same costume. Islam following christianity?

(12) GaEulKai wrote:
May 15th 2010 1:12 GMT
Being an American who has visited many Middle Eastern countries, I have seen western people just like me wear their dresses. I have seen women in bikinis soak the sun on the shores of the dead sea..However, I do think the only place where westerners are not permitted to dress the way they dress is in Saudia Arabia because that country is considered 'holy' by Muslims. Why can't everyone just be tolerant of one another and learn to live in peace? The human race seems disgusting now, who preys and hurts their own species? We do.

Finally, another recurrent feature which is worth mentioning concerns the recourse to extra-textual references used by the commenters both to show a sense of belonging to the epistemic community of *The Economist* and, again, to support the claims advanced, as shown in the following examples:

(13) Res Publica wrote:
May 13th 2010 5:02 GMT
--"As Voltaire might have said, "I disapprove of your dress, but I will defend to the death your right to wear it". True. And Montesquieu added: "My freedom ends where the others' freedom begins." In other words, respect and you will be respected. Is that that another "ethnocentric" western value??? Doesn't it apply the same way in Islam???

(14) Kim77 wrote:
May 13th 2010 8:59 GMT
There is an old Korean proverb that applies admirably to the case at hand. 'Scratching your sore spot will make it even sorer.' Is burqa an (eye)sore? For most people, yes. But banning it is probably the worst way to deal with it. You are giving burqas a justified victimhood status. No one can deny that opposition to burqas is motivated by xenophobia, thinly veiled with secularism and concern for female rights.

(15) zekk wrote:
May 18th 2010 10:35 GMT
"I disapprove of what you say, but I will defend to the death your right to say it." Voltaire never said that! This line comes from "The Friends of Voltaire" by Evelyn Hall (1907). Many people after reading the book thought Voltaire actually wrote or said that, but it is not the case. In 1935, Evelyn Hall declared: "I did not intend to imply that Voltaire used these words verbatim, and should be much surprised if they are found in any of his works"..Now I wouldn't be surprised that after reading The Economist, some might think that Voltaire actually said "I disapprove of your dress..." and will use this expression as Voltaire's.

The examples reported above corroborate the claim advanced in the present study that the language and rhetorical styles of commenters in this specific context display a high degree of variation ranging from a cline of formality/informality features which on the one hand bear the editor's rhetorical imprimatur and on the other hand reflect personal 'free' style traits typical of blogging.

6. Conclusion

The interactive exchanges between the editor and the commenters/readers analyzed in the selected sample display considerable variation in communicative styles. A general trend observed in the readers' responses is how they are conditioned by the editorial model at various levels. First, traits of editorial influence have been noted in the editor's tendency to orient reader (and commenter) topic saliency both by foregrounding the relevance of specific editorials and by selecting locally relevant leaders targeted to different audiences. The data reported in Table 1 display an overall correspondence between the rate of readers' responses and the editorials suggested by *The Economist* even if at times the contributors do not respond accordingly and select their preference independently from the editor's recommendations.

Second, another sign of editorial impingement resides in the editor's mandate regulating the flow of 'blogging' activity in this context. Indeed, as observed in Section 4, editorial norms alter significantly a few distinctive traits typically marking the generic integrity of the blog such as the structure organized in reverse chronological order and the recourse to continuous updates which, in this context, are limited temporally.

Third, the editor's complex rhetorical style is also reflected in the commenters' language. As noted in Section 4, *The Economist* has its own special style characterized by mixing formal/informal registers; alternating arguments and counterarguments in editorials; using interlocutory forms of direct address, often deliberately provocative, aimed at arousing the reader's interest; employing metaphorical constructs using sophisticated language, and creating obscure titles rich in extratextual references and puns targeted for a sophisticated audience.

Markers of the editorial 'proto-genre' in the commenters' language are particularly evident in the readers' consolidated trend to construct their own 'counter-editorials' by mimicking the editor's mixed communicative style and rhetorical strategies, as illustrated in the examples reported in Section 5.

The significant degree of influence of the editor's style observed in the commenters' stylistic repertoire brings to the fore the controversial issue of the blog's *generic integrity*. In the light of what has been observed it is in fact debatable whether the blog's *generic integrity* is determined on the basis of the recurrent formal features previously evidenced in the literature, as some of these recurrent features have undergone editorial imprinting, hence change, at least in this specific context.

The ephemeral quality of genres in web-mediated communication suggests that the unconventional definition of blogs proposed by Miller/Shepherd (2009: 283) deserves further consideration. Indeed, a blog can be: "a technology, a medium, a constellation of affordances – and not a genre".

References

Bhatia, Vijay K. 2004. *World of Written Discourse*. London/New York: Continuum.

Blood, Rebecca 2000. *Weblogs: A History and Perspective*. Available online at: <http://www.rebeccablood.net/essays/weblog_history.html>.

Blood, Rebecca 2003. *Weblogs and Journalism: do they connect?* Available online at: <http://www.nieman.harvard.edu/reports/article/101025/Weblogs-and-Journalism-Do-They-Connect.aspx>.

Bowker, Janet 2008. Lightening "The Dismal Science": The Dialogic Language Game in the Headlines of the Economist Magazine. *Textus* XXI/1, 167-186.

Crowston, Kevin / Williams, Marie 1997. *Reproduced and emergent genres*. Available online at: <http://crowston.syr.edu/system/files/7734060030-1.pdf>.

Devitt, Amy J. 2009. Re-fusing form in genre study. In Giltrow, Janet / Stein, Dieter (eds) *Genres in the Internet*. Amsterdam/Philadelphia: John Benjamins, 27-47.

Friedman, Thomas L. 2005. *The World is Flat*. London: Penguin Books.

Garzone, Giuliana 2007. Genres, multimodality and the World Wide Web: theoretical issues. In Garzone, Giuliana / Poncini, Gina / Catenaccio, Paola (eds) *Multimodality in Corporate Communication. Web genres and discursive identity*. Milano: Franco Angeli, 15-30.

Gillmor Dan, 2003. *Moving toward Participatory Journalism*. Available online at: <http://www.nieman.harvard.edu/reports/article/101062/ Moving-Toward-Participatory-Journalism.aspx>.

Giltrow, Janet / Stein, Dieter 2009. Genres in the Internet: Innovation, evolution, and genre theory. In Giltrow, Janet / Stein, Dieter (eds)

Genres in the Internet. Amsterdam & Philadelphia: John Benjamins, 1-25.

Gotti, Maurizio 2005. *Investigating Specialized Discourse*. Bern: Peter Lang.

Herring, Susan / Scheidt, Lois Ann / Bonus, Sabrina / Wright Elijah 2005. *Weblogs as a bridging genre*. Available online at: <http://ella.slis. indiana.edu/~herring/itp.pdf>.

Lemann, Nicholas 2006. The Wayward Press. Amateur Hours. Journalism without journalists. *The New Yorker,* August 7, 2006. Available online at: <http://www.newyorker.com/archive/2006/08/07/060807fa_fact>.

McNeill, Laurie 2005. Genre Under Construction: The Diary on the Internet. *Language@Internet* 2/1.

Miller, Carolyn R. 1984. Genre as social action. *Quarterly Journal of Speech* 70, 151-167.

Miller, Carolyn R. / Shepherd, Dawn 2009. Questions for genre theory from the blogosphere. In Giltrow, Janet / Stein, Dieter (eds) *Genres in the Internet.* Amsterdam & Philadelphia: John Benjamins, 263-290.

Mitchell, Bill 2003. Weblogs: *A Road Back to Basics*. Available online at: <http://www.nieman.harvard.edu/reports/article/101039/Weblogs-A-Road-Back-to-Basics.aspx>.

Myers, Greg 2010. *The Discourse of Blogs and Wikis*. London & New York: Continuum.

Regan, Tom 2003. Weblogs Threaten and Inform Traditional Journalism. Available online at: <http://www.nieman.harvard.edu/reports/article/101041/Weblogs-Threaten-and-Inform-Traditional-Journalism.aspx>.

Riley, Philip 2002. Epistemic Communities: The Social Knowledge System, Discourse and Identity. In Cortese, Giuseppina / Riley, Philip (eds) *Domain-specific English*. Bern: Peter Lang, 41-64.

The Economist, May 15th, 2010. "Banning the burqa. A bad idea... whose time may soon come in parts of Europe". Available online at: <http://www.economist.com/node/16108394>.

Wenger, Etienne 1998. *Communities of Practice. Learning, Meaning, and Identity.* Cambridge: Cambridge University Press.

Małgorzata Sokół

Metadiscourse and the Construction of the Author's Voices in the Blogosphere: Academic Weblogs as a Form of Self-promotion

1. Introduction

Academic[1] weblogs (or blogs) are a relatively recent phenomenon of scholarly exchange, offering opportunities for increased academic interaction and self-expression outside the traditional scholarly publishing system. Engagement in blogging activities by academics, as well as their growing, multiple participation in other online environments, has reinforced changes in the contemporary world of academia. The emergence of new, largely informal, web-based modes of scholarly communication and innovative work and research tools have been accompanied by the development of new forms of scholarly publication and knowledge storage and alterations in work and research patterns. The intensity of these changes may be indicative of a deeper transformation on the institutional level, where well-established scholarly norms and discursive practices are being challenged. The technology-driven changes are also reflected in patterns of socialisation into the academic world and gaining professional status and recognition.

The aim of this chapter is to analyse the function of weblogs in scholarly communication through the investigation of authoring practices of blogging members of the academia. More specifically, I will focus on how academic bloggers negotiate their (professional) identity through engagement expressed in the use of metadiscoursal resources of self-mention and citation. The discussion will be supported by the

1 Other names used in reference to blogs run by academics and for academic purposes include *science*, *research* or *scholarly blogs*.

analysis of blog entries that come from eight academic weblogs (for details, cf. Appendix). In general, I will argue that academic weblogs function as platforms for extensive self-expression and self-promotion under the generally accepted, 'traditional' norms of scholarly exchange.

2. Web 2.0 as a collaborative technology

The Internet functions primarily as a social technology, as evidenced by its current development, commonly referred to as Web 2.0 or the Social Web. Essentially, Web 2.0 encompasses a second generation of Web services and applications that facilitate interactive information sharing, user-centred design, and more dynamic user participation and collaboration (cf. e.g. O'Reilly 2005; Harrison/Barthel 2009). Apart from blogs, examples of Web 2.0 include social networking sites, video sharing sites, wikis, mashups and folksonomies.[2]

Although there is some lack of clarity as to what precisely the concept of Web 2.0 refers to, it is acknowledged that the foundation of the current stage of web development is the transformed role of the medium's audience: from mere passive web consumers to active creators of content that is willingly shared with others. While several studies have pointed out that the idea of an active and creative audience is neither historically new nor confined to the development of new information and communications technologies (Harrison/Barthel 2009; Zimmer 2009), from the very beginning of the widespread development of the web, users noticed and exploited its collaborative and participatory potential. However, as Harrison and Barthel emphasise, "the popularity of Web 2.0 applications demonstrates that, regardless of

2 *Wikis* are community publishing tools or websites that allow users to edit and control content. *Mashups* are websites or applications that combine content from two or more sources into a new single integrated form. *Folksonomies* are user-generated taxonomies used to categorise and retrieve information on the Internet (cf. the Glossary in Sonwilla-Weiss 2010: 246-252).

their technical expertise, users can wield technologies in more active ways than had been apparent previously" (2009: 157).

This active engagement of users is primarily enabled by the inherent technological characteristics of Web 2.0 applications and services that contribute to the construction of an 'architecture of participation' and make it possible to 'harness collective intelligence' (O'Reilly 2005). In his discussion of the phenomenon of Web 2.0, O'Reilly draws attention to the relevance of hyperlinking as the foundation of web existence, thanks to which users discover content and link to it. Consequently, numerous Web 2.0 services and portals thrive as a result of the intensive, repetitive and collective activity of their users.

An important point is that users' engagement in generating content does not require advanced technical skills (O'Reilly 2005; Harrison/Barthel 2009). Users can create and edit web pages without the knowledge of HTML[3] and web design programmes. And although the motivations behind their active participation in web design are still insufficiently understood, users' willingness most probably results from the high gratification they receive from their activity.

In addition, 'social' applications and services are attractive because on the one hand, they support collaboration and group interaction without imposing structure or organisation (cf. Schiltz *et al.* 2007), but on the other, they also enable users to customise their content according to their individual needs and purposes, including the realisation of commercial or promotional aims (Harrison/Barthel 2009). In the collaborative process of content production, it is the profile of an individual user which is considered an essential commodity of Web 2.0. Thus, the web today is still a hybrid communicative context, in which the individual develops alongside the collective, the private merges with the public and the institutional, or the commercial coexists with the 'democratic' and free access.

The web has always been a dominant technology for the collection, organisation and retrieval of knowledge and information (O'Reilly 2005; Zimmer 2009). Web 2.0, with its collaborative content generation, provided the information tools which have contributed to further transforming the ways in which knowledge is structured and

3 HTML stands for HyperText Markup Language used to build web pages.

stored (note the term *infoware* used with reference to Web 2.0 software for producing and sharing content; O'Reilly 2005). The hyperlinking inherent in the Social Web provides a novel means of accessing, cross-referencing and verifying information.

3. Social software and the transformation of academia

The technological features of Web 2.0 referred to above bear significance for the academic community and the multiple ways in which it adopted the new medium. With the emergence of Web 2.0, the academic community started a new phase of web exploitation, and thus entered a new phase of its own transformation. It seems, in particular, that the reconceptualisation of the user in the context of content generation afforded by the Social Web has important implications for academia, as being an active knowledge creator has always been essential for building professional status.

The research on current trends in scholarly communication (e.g. Merz 2006; Maron/Smith 2009; Sokół 2012) indicates that next to the established 'traditional' forms of exchange and publishing that still prevail, academics take advantage of numerous digital resources. Their decentralised distribution and direct accessibility mean that digital resources significantly contribute to the process of scholarly communication, leading to changes in academia at the organisational and institutional level.

According to the report by Maron and Smith (2009), popular web-based resources for scholarly communication include e-only journals, digital reviews, preprints and working papers, encyclopaedias and dictionaries, data resources, blogs, discussion forums and hubs such as portals for scholarly organisations and societies. These resources differ as to the degree of the 'social' or participatory element in them. However, more and more of these undertakings are developed through the incorporation of multimedia technology and networking tools to enhance their interactive potential and invite more audience engagement. In general, academics use the digital environment: to accelerate the speed of publication and peer-review process; to share and access new research, either the final data outputs of particular scientific undertakings or their interim results; to exchange news about relevant academic endeavours; and to stay in touch with a broader community of scholars and engage in informal discussions with them. Although their research does not focus on statistical analyses, Maron and Smith note that despite the fact that digital formats are more convenient for adoption by some disciplines, examples of digital resources can be found across all disciplines (2009: 19). What is also important from the perspective of the present research, Maron and Smith's investigation reveals the impact of traditions of scholarly culture upon digital publishing. As establishing scholarly legitimacy is essential in academia, it is common for digital resources to incorporate (open) peer review and editorial check. In addition, the same purpose is accomplished in the digital environment through credentialling and citation metrics. The relation between epistemic cultures and practices and patterns of adoption of digital technology is confirmed by other researchers: cf. e.g. the ethnographic case study of email-based interaction and pre-printing practices in the area of theoretical particle physics by Merz (2006), or the genre analyses of scholarly blogs in physics by Kjellberg (2009) and of an e-forum in digital humanities by Sokół (2012).

4. Weblogs as a new genre of scholarly exchange

The blogosphere is rapidly becoming an important space for scholarly exchange (e.g. Bonetta 2007; Dennen 2009; Kjellberg 2009; Mortensen/Walker 2002), largely due to the technological characteristics of weblogs. In general, weblogs are considered to be the next stage of development of personal home pages. They function as frequently updated diaries, in which postings are presented in a reverse chronological order and access to them is open to the public (Herring *et al.* 2004; Schmidt

2007). With respect to their structure, weblogs are characterised by a large degree of diversity, but the elements common to all are comment sections, archives and sidebars. Since an essential feature of weblogs is linking, the technologies responsible for the rise of blogging are considered to be RSS (Really Simple Syndication), a format for syndicating content on sites and used to push all kinds of data updates, and the permalink that essentially enables effective linking between weblogs. The combination of RSS and permalinks affords functionalities that are greatly appreciated by the growing community of bloggers:

> Not only can people subscribe to each others' sites, and easily link to individual comments on a page, but also via a mechanism known as trackbacks, they can see when anyone else links to their pages, and can respond, either with reciprocal links, or by adding comments. (O'Reilly 2005)

In effect, the blogosphere has developed as a potent, densely-interlinked network within the Social Web, with the collective power to filter content and shape search engine results. The blogosphere is a self-referential community, whose members pay attention to other bloggers and blogs, and thus strengthen their power and visibility.

The popularity of weblogs is also linked to the development of blogging software, which is freely available and makes starting a blog a fairly straightforward task, even for users with low technical skills (Bukvova *et al.* 2010; Mortensen/Walker 2002). Up-to-date estimations as to the number of blogs and detailed statistical analyses of the blogosphere are possible thanks to blog search engines such as Technorati <www.technorati.com> or systems for indexing links on weblogs[4]. As regards the use of weblogs by academia, Technorati.com currently indicates over 6,600 blogs with the tag *science* (the number as of February 2011). However, as Bonetta (2007) points out, the search engine results do not reveal the real number of academic blogs. For instance, according to the findings she reports, the actual number of academic blogs in 2007 was between 1,000 and 1,200 (Bonetta 2007: 443).

4 An example of such a system was Blogdex run in the years 2001-2006 by the MIT Media Lab in Boston. For a discussion of the system cf. e.g. 'Tracking Bloggers with Blogdex' by Leander Kahney, *Wired*. June 30, 2001 or Mortensen/Walker 2002.

From a technological perspective, weblogs are well investigated, but there are relatively few linguistic analyses of the genre (exceptions include Grieve *et al.* 2010; Herring *et al.* 2004; Herring/Paolillo 2006), let alone of academic weblogs (for an example of a genre-based analysis cf. Kjellberg 2009). What emerges from the basically ethnographic research is that the academic weblog is still a new and evolving genre, characterised by a large degree of diversity as to form, content and function (e.g. Bukvova *et al.* 2010; Kouper 2010; Mortensen/Walker 2002). Academic weblogs are considered a hybrid between journal, academic publishing, storage space for links and site for academic discussion, which potentially extends beyond the academic community (Mortensen/Walker 2002; also Kjellberg 2009). On the one hand, by their diary-like nature, academic blogs serve as media for self-expression and self-reflection. On the other hand, as social software, they have the capacity to engage others in collaborative activity, scrutiny, knowledge sharing and debate (Dennen 2009; Ewins 2005; Kjellberg 2009; Mortensen/Walker 2002), and are increasingly used as an effective teaching tool (e.g. Wheeler 2010).

Given the variety of purposes for which academic weblogs are used, they could be categorised as *knowledge blogs* (i.e. *k-logs*) or *mixed-type blogs* according to the classification of blogs suggested by Herring *et al.* (2004). In general, single posts in a blog are always written in relation to another post or text outside the blog or the blogger's experience, which makes them both subjective and specific at the same time.

The networking and sharing are accomplished within the academic community, but academic weblogs also have the power to reach lay audiences and engage them in scientific topics. Academic bloggers find the idea of open science attractive and frequently in their posts mix science with non-scientific content to facilitate the dissemination of knowledge and to encourage discussions about science (Kouper 2010). This kind of interaction in academic blogs results in changed author/audience relations and a merged private vs. public perspective. Academic blogs make it possible for individuals to cultivate their autonomous voice, but also realise their democratising potential through making science open to the broad public, at least potentially (Mortensen/Walker 2002).

In addition, academic weblogs are considered to serve as research tools and publication platforms (Bonetta 2007; Bukvova *et al.* 2010; Ewins 2005; Kjellberg 2009; Mortensen/Walker 2002). The self-reflection that is possible thanks to the specificity of the communication medium facilitates the focus on research and the analysis of bloggers' own thoughts:

> Writing in a weblog one is forced to confront one's own writing and opinions and to see them reflected in the words of others. The discussions are much more open and also more permanent than discussions in a seminar room or at a conference. A blog is a permanent archive (as long as the writer preserves the archives and the server remains online) and it is searchable. What you write in your blog can be quoted and discussed in any forum. (Mortensen/Walker 2002: 269)

Treating weblogs as a convenient and speedy publication space is increasingly common among academic bloggers, but definitely the practice raises an important issue concerning scholarly norms and practices. Blogging by academics is sometimes considered as part of popular culture or as folk journalism, and thus can be a challenge to traditional publishing practices (Mortensen/Walker 2002: 252). Blog posts are less carefully constructed and lack the refinement expected from traditional publishing formats (Ewins 2005: 370). Moreover, as writing about one's research or ideas in a blog is in fact a form of announcing them, publishing them through traditional channels can be more difficult and there is a danger that ideas will be stolen. Still, the announcement of one's research on a blog means that its ownership can be claimed. Besides, publishing online is quicker and less restricted in terms of users' scholarly status and expertise (Mortensen/Walker 2002).

Notably, as academic blogs are research and publication tools, they constitute a powerful resource for constructing an online professional identity (e.g. Bukvova e*t al.* 2010; Dennen 2009; Ewins 2005; Kirkup 2010; Kouper 2010). To quote an academic blogger's words:

> I use my blog as an online log of what seem to me the most important readings, concepts and themes from the course; and as a reflective journal which helps me to bed down new knowledge, consolidate my thoughts about what I've read, process my thinking, and make connections between ideas […].

> Often in my posts I'm trying on new concepts for size, testing their fit with existing ideas, trying out a new terminology to see what it feels like in use, and more or less consciously acting out an academic persona – performing my own identity as a practitioner of the academic craft. (B7, April 22, 2010)

While setting up a blog may be an easy task, establishing an accepted online presence is a long-term process requiring on the part of the blogger regularity, persistence and simply time. Still, this consistency is essential in legitimating the online researcher (Mortensen/Walker 2002: 250-251), adding an element of stability to the changeable context of the weblog. Moreover, regular updating of the weblog not only creates a strong sense of the author's presence, but also a sense of community among people with similar interests (Ewins 2005: 369).

Engagement in blogging practices has been on the increase among members of academia, with some universities attempting to establish blogging platforms on their servers.[5] However, academic blogging also evokes negative attitudes and raises several concerns, primarily related to privacy and job security issues, and academic freedom. There are reports of institutional conflicts that resulted from the bloggers' unfavorable discussion of their universities, or situations when the mere fact of running a blog turned out to be a serious impediment to the blogger's career (e.g. Bonetta 2007; Dennen 2009; Ewins 2005).

5. Constructing scholarly identity in weblogs through metadiscourse: Analysis and discussion

The focus of this chapter is the authoring practices of members of academia in their weblogs. More specifically, my aim is to analyse how bloggers negotiate their (professional) identity through engagement expressed in the use of metadiscourse. Although my main concern is

---

5 An example is the University of Warwick, UK. See the article 'By the Blog: Academics Tread Carefully" by Zoe Corbyn, *Times Higher Education*, 9 October 2008, available at: <http://www.timeshighereducation.co.uk/story.asp?storycode=403827>.

blogger's verbal expression, I will also refer to visual and structural elements of the blog relevant for identity construction, as blogs tend to be increasingly multimodal. The theoretical-methodological basis of the analysis is an integration of a discursive approach to identity construction (Benwell/Stokoe 2006; Bucholtz/Hall 2005) and van Dijk's (2008) sociocognitive model of discourse analysis. In this way, I emphasise that an individual's identity is dynamically negotiated in discourse in relation to the social structures that an individual is a member of and his/her social roles. In addition, as users' discursive choices may be explained in terms of their individual needs and purposes, and their subjective understanding of the communicative situation, I draw attention to an individual's active role in identity construction.

For the content analysis of the research material, I adapt Hyland's model of metadiscourse (Hyland 2005a) and focus on the use of resources of self-mention and citation (evidentials). Hyland's work represents an integrative view of metadiscourse, which accommodates the textual and interpersonal function of metadiscoursal signals, and therefore it can offer a complete account of the author's presence in a text (cf. also Hyland 2005b). Resources of self-mention refer to the degree of explicit authorial presence in the text. Evidentials provide the source of knowledge and their role is to guide the readers' interpretation, and to establish an authorial command of the subject. In this way, I look into how bloggers explicitly mark their presence in their posts through self-reference, and how they strengthen their professional authority and credibility through attribution to other (reliable) sources of information. In addition, this approach allows me to investigate both individual and community-related aspects of academics' online identity in the weblog.

As the genre of academic blogs is considered a hybrid genre and exemplifies a large degree of discipline-bound variation, I have narrowed my analysis to the English blogs of humanities scholars. The research material used here consists of posts retrieved from eight single-author blogs of humanities scholars (mainly of linguistics and philosophy), who run their blogs fairly regularly under their real name (three females and five males; for references to the weblogs sites, cf. Appendix). The bloggers' real life identities are revealed either in the blog's name or the blogger's profile. The posts used in the present study come from the bloggers' activity between November 2009 and April 2010 and the total size of the corpus amounts to 111,903 words. The qualitative analysis of the corpus was accompanied by quantitative investigations, conducted with the support of the concordancing software *Text Analyser.*[6]

5.1. Self-mention

Self mention denotes resources that explicitly indicate authors' presence in the text (Hyland 2002; 2005a). On the level of verbal expression, self-mention is essentially measured by the frequency of first-person pronouns and possessive adjectives. Pronoun choice is considered an important linguistic device that can be interpretable in terms of power relations, and has received much attention from researchers concerned with analyses of academic discourse (e.g. cf. Hyland 2002; 2005a and 2005b for the analyses of written academic registers; Fortanet 2004 and Rowley-Jolivet/Carter-Thomas 2005 for the analyses of spoken academic genres; Sokół 2012 for the analysis of electronic academic discourse). Several studies reveal the extensive presence of first-person pronouns in a variety of contexts of academic exchange, indicating the progressive personalization of academic discourse. Skilful positioning by means of pronouns enhances the promotion of the author's scholarly competence, authority and acceptance of their work, necessitated by the growing competition within academia and the commodification of scholarship and identity (Fairclough 1995). The use of self-mention resources indicates the author's conscious decision to emphasise an individual perspective and to take personal responsibility for his/her claims, which enhances the author's professional status and expertise.

The analysis of bloggers' presence in the corpus under investigation focused on the realisations of self-mention on the levels of both verbal and visual expression. The analysis of the distribution of first-person pronouns in the data (cf. Table 1) has been extended to include

6 *Text Analyser* is a non-commercial concordancing software for Windows, designed for my personal use in the corpora-based research of web genres.

the referential meanings of the pronoun *we* (cf. Table 2). Apart from checking the ratio of the *I*- versus *we*-perspective, it was interesting to investigate collective aspects of the blogger's identity. Because of its relevance for studying author-audience relations, Table 1 additionally includes the figures for the pronoun *you*.

Self-reference resources	*Occurrences*	*Number per 1,000 words*
I	1,004	8.97
me	114	1.02
my	285	2.55
we	556	4.97
us	98	0.87
our	158	1.41
you	422	3.77

Table 1. Number of occurrences of self-mention resources and of the pronoun *you*.

The frequency of *I* in relation to *we* in the data under investigation is almost twice as high, which indicates the bloggers' need to mark their authorial presence. This is certainly due to the self-reflective nature of academic weblogs. Still, the prominence of *I* may also indicate the bloggers' active and responsible participation in discussions, self-promotion and propagation of their research. Interestingly, the frequency of *I* in academic weblogs is lower than in e-seminars (ca. 12-14 occurrences of *I* per 1,000 words, Sokół 2012), but in general comparable with the findings for spoken academic registers (cf. e.g. Fortanet 2004; Rowley-Jolivet/Carter-Thomas 2005). As regards the pronoun *we*, its distribution is markedly lower than in spoken academic registers, but approximates the figures for e-seminars.

Particular uses of the pronoun *I* reflect the variety of functions of academic weblogs related to issues of identity construction. The bloggers' expression in the entries under study is self-oriented and blurs the boundaries between the private and the public. Apart from personal narratives or comments (cf. examples (1)-(3)), the entries may include information about the bloggers' engagement in academic undertakings, especially in the offline world (examples (4)-(6)), as well as announcements and discussions of their own research ideas and findings (examples (7) and (8)).

(1) They write to me because I was foolhardy enough fifteen years ago to launch a website on the topic of EE, with the aim of bringing together the material on the subject that I could find on line or, in most cases, put on line. (B8, April 8, 2010)

(2) When I first saw the top of this cork I couldn't help but sing 'Tiptoe Through the Tulips'. I guess this is because verbalizing the words results in the realization of 'tip toad' as the homophonic past tense 'tiptoed'. (B4, January 20, 2010)

(3) According to the federal government, which is never wrong about these things, I myself became officially old, and therefore useless as a repository of wisdom and memory, last Spring. (B6, November 9, 2009)

(4) Editors still remind me that I mustn't aim my writing at a bunch of stuffy academics. (B6, February 10, 2010)

(5) Quite often I receive emails from high school students writing papers about blogging who have long lists of questions they'd like me to answer. (B7, January 12, 2010)

(6) I have a student in a philosophy of biology class who is about to transfer into a wine-making program. (B3, January 15, 2010)

(7) In my little study of Wikipedia, I initially stumbled on the difference between featured and regular articles. If I had thought about it in advance, I would not have tested any featured articles at all. I had included them, however, so I reported the results and suggested that the data about featured articles be set aside. (B2, April 15, 2010)

(8) When I say 'the door is locked', I first and foremost utter a declarative sentence. Whether I also made an assertion depends in part on the social norms that come in force. (B5, April 2, 2010)

Thus, academic weblogs serve as information boards thanks to which academics can follow developments in a given field, network with fellow researchers and exchange ideas with them. At the same time, being treated by their authors as research and publication tools, academic weblogs also serve to realise self-promotional objectives (for examples, see (9) and (10)).

(9) I am now committed to drawing up a new paper on advice on reviewing papers that I will post on this blog probably in the next month or two. My thanks

again to the committee for kindly inviting me to take part in this session. (B1, December 31, 2009)

(10) I did not use the term 'pre-fortis clipping' in Accents of English (1982) which I presumably would have done had it been available then – but I did use it in the first edition of LPD (1990), as well as in my article 'Syllabification and Allophony' published in Susan Ramsaran (ed.), Studies in the pronunciation of English, A commemorative volume in honour of A.C. Gimson (London and New York: Routledge, 1990), pages 76-86, where I said of Pre-fortis clipping. (B8, April 21, 2010)

The frequency analysis of the referential meanings of the pronoun *we* provides additional information about the patterns of academic identity construction in weblogs (cf. Table 2). In my data, I identify three uses of the pronoun *we*:

- 'general *we*' – used to refer to both 'exclusive *we*' and '*we*-the human race' (cf. Vassileva 2000), as the context in the concordances was insufficient to distinguish between the two; besides, the use of *we* without explicit referents is common among members of the academia; it may also be the case that in such cases the referent is, rather unconsciously, taken for granted by an academic, or as Biber *et al.* point out, authors may inconsistently switch between different uses of *we* (Biber *et al.* 1999: 329);
- '*we*-represented academic institution'; this category includes the reference to the author and his/her academic institution, department or institute, immediate colleagues and research partners, students, etc.;
- '*we*-blogging community; this category of 'inclusive *we*' is relevant for the discussion of the sense of group identity among the bloggers under investigation and their relationship with the (potential) audience.

Referential meanings of we	*Occurrences*	*%*
General *we*	244	43.88
we-academic community	216	38.85
we-blogging community	96	17.27

Table 2. Number of occurrences of referential meanings of the pronoun *we*.

It has to be noted here that, since sub-categorising the different uses of *we* revealed ambiguous cases, attribution to a given category was not always a straightforward task. Generally, though, it can be argued that individual bloggers seem to enhance their professional status through the expression of a strong sense of membership in the general academic community (high frequency of the '*we*-academic community': see examples (11)-(13)). At the same time, occurrences of the '*we*-blogging community' may indicate the sense of community in relation to other blogging members of academia and the blogger's wish to engage them in ongoing discussions (see examples (14)-(16)). The use of 'general *we*' is similar to the patterns found in spoken and written discourse.

(11) We've received nearly a million euros to fund a large, European project exploring creativity in the field of electronic literature. (B7, January 12, 2010)

(12) We have two PhD fellowships advertised right now, with a deadline of January 31. (B7, January 5, 2010)

(13) I teach at a department with an unranked program, and we place the majority of our graduates in tenure track jobs. (B2, March 25, 2010)

(14) [...] and we have begun to see that in another topic we have recently debated. (B1, February 27, 2010)

(15) As long as we have been on the subject of lists this week, here is a list of the top ten posts from 2009 on A Walk in the Words. (B4, January 15, 2010)

(16) I treat the items we are discussing as basically sequences of a strong vowel plus a weak vowel. (B8, December 10, 2009)

The low frequency of the pronoun *you* in the corpus under study reveals bloggers' relatively low awareness of the audience. This may also result from the higher prominence of the individual perspective and the focus on self-expression that is characteristic of the genre of academic weblogs. Still, the presence of the pronoun *you* in the data may indicate spontaneity and indirectness in addressing the readers, and overall, a less formal tone (for examples, see (17)-(19)).

(17) A question for readers: who would you like to see as the next philosopher on tv? (B1, April 02, 2010)

(18) To give you an idea of how popular this line is, when I went to google the phrase "monkey-fighting snakes" [...]. (B4, February 15, 2010)

(19) Suppose tonight you will fission into two persons. One of your successors will wake up on Mars and one on Venus. There are then two possibilities for how things might be for you tomorrow: you might wake up on Mars, and you might wake up on Venus. (B5, March 26, 2010)

In addition to self-mention expressed through pronoun choice, the study reveals numerous occurrences of self-reference realised multimodally. The examples noted in the data include videos with the bloggers' talks and conference presentations, their Power Point presentations, links to calls for papers and conference announcements in the organisation of which bloggers are actively engaged, links to word documents with paper drafts, graphical images of books authored by the bloggers, with links to online bookshops or reviews of these books, and finally, private photographs. These elements function as announcements of the bloggers' ideas and research, even if in its interim stages, promote their scholarly undertakings, and most frequently, serve as starting points of their posts. This variety of multimodal resources of self-mention and their functions shows that blogging academics take advantage of the technical possibilities offered by the medium to facilitate effective scholarly exchange, and appropriate the medium to suit their purposes of self-promotion and self-empowerment.

5.2. Citation

Citation or evidentials refer to resources that indicate the source of knowledge. While the resources of self-mention belong more to the interpersonal level of discourse organisation, citations are primarily used to organise propositional content in the text. Their role is to "guide the reader's interpretation and [to] establish an authorial command of the subject" (Hyland 2005a: 51). As Hyland also argues, references to community-based literature generally enhance the persuasive impact of the text: they provide support for arguments, display an allegiance to a particular community, and establish a credible author voice (2005a: 56). Citation marks authors' attempts to negotiate propositional information in ways that are meaningful and appropriate to a particular disciplinary community. This emphasises the relevance of the collaborative construction of knowledge:

> The embedding of arguments in networks of references not only suggests an appropriate disciplinary orientation, but also reminds us that statements are invariably a response to previous statements and are themselves available for further statements by others. New work has to be embedded in the literature of the community. (Hyland 2005a: 158)

The study of the citation practices in the corpus indicates that the bloggers under investigation consciously construct their identity of competent, trustworthy and authoritative scholars. Being aware of the possible consequences of violation of the accepted norms, and having too much to lose, the bloggers follow the rules of proper academic conduct which they transfer from the offline world.

On the verbal level, the citation resources identified in the corpus do not differ much from the ones employed in the traditional, more formal genres of scholarly communication. Among the most frequent are adverbials beginning with *As* (e.g. *As reported here by...*, *As recounted in the book...*, *As described in the conference announcement...*, *As X puts it/says, argues...*, *As noted in...*, *As predicted by....*, *As Lewis argued in 1973...*), the use of the complex preposition *according to*, and most frequently, pointing directly to the source of information indicating the name of the quoted person, the title of the book or simply the name of the resource referred to. The new aspects of verbal citations in the weblogs are definitely a lower degree of formality, expressed for instance through more direct references to the quoted individuals or a certain degree of familiarity and intimacy (e.g. *As John puts it...*), and the presence of medium-related references (e.g. *As Wikipedia tells us...*, *As posted by....*, *As commented by...*).

The novel patterns of citation are also revealed through the analysis of the visual aspect of the weblogs. Academic referencing is frequently realised through the hypertextual links to the referred sources or the incorporation of the visual and audiovisual materials referred to (e.g. dictionary entries, videos, book covers, blog entries). An example (cf. Figure 1) includes an image of another author's book referred to in the post by B8. On the surface, citations realised multimodally may appear random and casual, but in fact, as Mortensen/Walker point out (2002: 261), they follow the acknowledged norms and practices pretty rigorously.

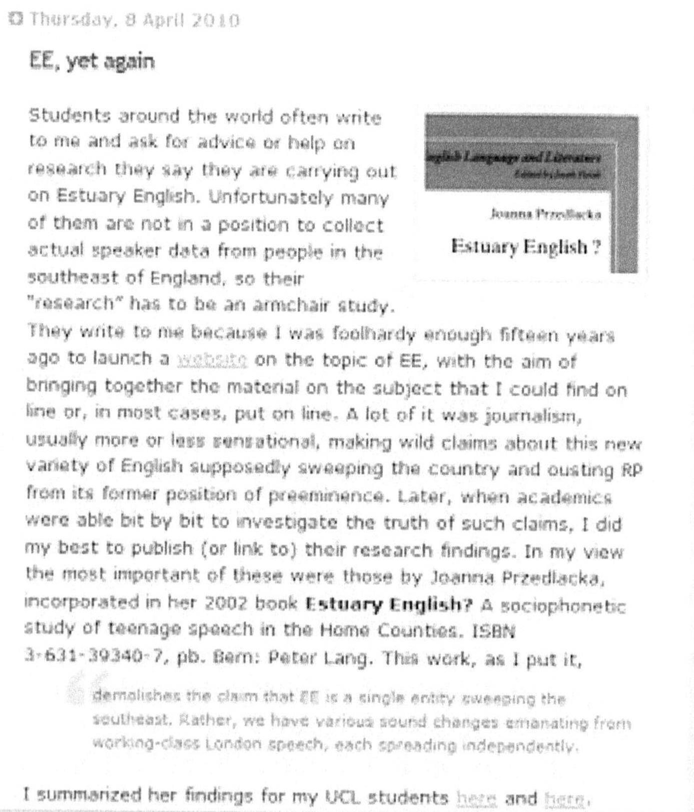
Thursday, 8 April 2010

EE, yet again

Joanna Przedlacka

Estuary English ?

Students around the world often write to me and ask for advice or help on research they say they are carrying out on Estuary English. Unfortunately many of them are not in a position to collect actual speaker data from people in the southeast of England, so their "research" has to be an armchair study. They write to me because I was foolhardy enough fifteen years ago to launch a website on the topic of EE, with the aim of bringing together the material on the subject that I could find on line or, in most cases, put on line. A lot of it was journalism, usually more or less sensational, making wild claims about this new variety of English supposedly sweeping the country and ousting RP from its former position of preeminence. Later, when academics were able bit by bit to investigate the truth of such claims, I did my best to publish (or link to) their research findings. In my view the most important of these were those by Joanna Przedlacka, incorporated in her 2002 book **Estuary English?** A sociophonetic study of teenage speech in the Home Counties. ISBN 3-631-39340-7, pb. Bern: Peter Lang. This work, as I put it,

> demolishes the claim that EE is a single entity sweeping the southeast. Rather, we have various sound changes emanating from working-class London speech, each spreading independently.

I summarized her findings for my UCL students here and here.

Figure 1. The screenshot from B8's blog entry that includes an example of citation realised multimodally.

6. Conclusions

In general, this research has provided evidence that academic identity as constructed in the weblog, similarly to other asynchronous types of Internet situations (e.g. e-forum), is an extension of an academic's real life identity, and that interaction in the weblog is largely governed by the scholarly rules and norms transferred from the offline world. The analysis of the data reveals an extensive use of resources of self-mention, which confirms that blogs represent a very self-oriented genre, used mainly for the blogger's self-expression and self-reflection, but also for self-promotion. At the same time, the distribution and types of citation resources indicate the blogger's adherence to the established norms of the scholarly community. Bloggers exploit the possibilities of the new medium and increasingly rely on multimodal resources to effectively negotiate their identity of a competent and reliable scholar. Consequently, academic weblogs are a communicative situation in which public, institutional and private aspects merge.

Weblogs are increasingly used as publication and research platforms, which is on the one hand enabled by their technological format, but on the other hand, is necessitated by the commodification of science, and the speed and immediacy of developments in the competitive and diversified academic world. Regular and reliable engagement in a blogging activity may to some extent enhance the bloggers' career development and the propagation of their research and academic achievements. Given the degree of explicit self-reference and bloggers' adherence to the established norms of scholarly citation, it seems that academic weblogs can be treated as a new form of self-publishing. In academic weblogs, most choices as regards identity work depend upon the individual blogger; verification and evaluation, however, come from the blogging community. In the weblog, peer-review proceeds largely through comment sections and interlinking, and it is up to the general academic community whether such a form of publishing will gain more formal recognition.

The results of the research indicate the emergence of less constrained modes of academic exchange, which follows from the current pursuit of open science, where weblogs have the potential to bring science to the public. The study suggests that the author's presence as marked on a blog is multi-layered and polyphonic, and challenges the established practices of scholarly communication. Further insights into the role of weblogs in contemporary scholarly exchange can be gained by more extensive analyses of metadiscoursal and evaluative resources to take into account discipline-bound variation and the subgeneric diversity within the genre of the academic weblog.

References

Benwell, Bethan / Stokoe, Elizabeth 2006. *Discourse and Identity.* Edinburgh: Edinburgh University Press.

Biber, Douglas / Johansson, Stig / Leech, Geoffrey / Conrad, Susan / Finegan, Edward 1999. *Longman Grammar of Spoken and Written English.* Harlow: Longman.

Bonetta, Laura 2007. Scientists Enter the Blogosphere. *Cell* 129/3, 443-445.

Bucholtz, Mary / Hall, Kira 2005. Identity and Interaction: A Sociocultural Linguistic Approach. *Discourse Studies* 7/4-5, 585-614.

Bukvova, Helena / Kalb, Hendrik / Schoop, Eric 2010. What We Blog? A Qualitative Analysis of Researchers' Weblogs. In *Publishing in the Networked World: Transforming the Nature of Communication*, 14th International Conference on Electronic Publishing 16-18 June 2010, Helsinki, Finland, 89-97. <http://dhanken.shh.fi/dspace/bitstream/10227/599/9/8bukvova_kalb_schoop.pdf>.

Corbyn, Zoe 2008. By the Blog: Academics Tread Carefully. *Times Higher Education.* Available at <http://www.timeshighereducation.co.uk/story.asp?storycode=403827>.

Dennen, Vanessa Paz 2009. Constructing Academic Alter-egos: Identity Issues in a Blog-based Community. *Identity in the Information Society* 2/1, 23-38.

Ewins, Rory 2005. Who Are You? Weblogs and Academic Identity. *E-Learning and Digital Media* 2/4, 368-377.

Fairclough, Norman 1995. *Critical Discourse Analysis: The Critical Study of Language*. London / New York: Longman.

Fortanet, Inmaculada 2004. The Use of 'We' in University Lectures: Reference and Function. *English for Specific Purposes* 23/1, 45-66.

Grieve, Jack / Biber, Douglas / Friginal, Eric / Nekrasova, Tatiana 2010. Variation Among Blogs: A Multi-Dimensional Analysis. In Mehler, Alexander / Sharoff, Serge / Santini, Marina (eds) *Genres on the Web. Computational Models and Empirical Studies*. Berlin: Springer, 303-322.

Harrison, Teresa M. / Barthel, Brea 2009. Wielding New Media in Web 2.0: Exploring the History of Engagement with the Collaborative Construction of Media Products. *New Media & Society* 11/1-2, 155-178.

Herring, Susan C. / Scheidt, Lois Ann / Bonus, Sabrina / Wright, Elijah 2004. Bridging the Gap: A Genre Analysis of Weblogs. In *Proceedings of the 37th Hawaii International Conference on System Sciences (HICSS'04)*. Los Alamitos: IEEE Computer Society Press.

Herring, Susan C. / Paolillo John C. 2006. Gender and Genre Variation in Weblogs. *Journal of Sociolinguistics* 10/4, 439-459.

Hyland, Ken 2002. Authority and Invisibility: Authorial Identity in Academic Writing. *Journal of Pragmatics* 34/8, 1091-1112.

Hyland, Ken 2005a. *Metadiscourse. Exploring Interaction in Writing*. London / New York: Continuum.

Hyland, Ken 2005b. Stance and Engagement: A Model of Interaction in Academic Discourse. *Discourse Studies* 7/2, 173-192.

Kahney, Leander 2001. Tracking Bloggers with Blogdex. Wired. Available at <http://www.wired.com/culture/lifestyle/news/2001/07/45546?currentPage=all>.

Kirkup, Gill 2010. Academic Blogging: Academic Practice and Academic Identity. *London Review of Education* 8/1, 75-84.

Kjellberg, Sara 2009. Scholarly Blogging Practice as Situated Genre: An Analytical Framework Based on Genre Theory. *Information Research* 14/3, Paper 410. <http://InformationR.net/ir/14-3/paper410.html>.

Kouper, Inna 2010. Science Blogs and Public Engagement with Science: Practices, Challenges, and Opportunities. *Journal of Science Communication* 9/1. <http://jcom.sissa.it/archive/09/01/Jcom0901%282010%29A02/Jcom0901%282010%29A02.pdf>.

Maron, Nancy L. / Smith, K. Kirby 2009. Digital Scholarly Communication: A Snapshot of Current Trends. *Research Library Issues: A Bimonthly Report from ARL, CNI, and SPARC* 263, 10-20. <http://www.arl.org/resources/pubs/rli/archive/rli263.shtml>.

Merz, Martina 2006. Embedding Digital Infrastructure in Epistemic Culture. In Hine, Christine M. (ed.) *New Infrastructures for Knowledge Production: Understanding E-Science*. London: Information Science Publishing, 99-119.

Mortensen, Torill / Walker, Jill 2002. Blogging Thoughts: Personal Publication as an Online Research Tool. In Morrison, Andrew (ed.) *Researching ICTs in Context*. Oslo: InterMedia, University of Oslo, 249-279.

O'Reilly, Tom 2005. What Is Web 2.0. Design Patterns and Business Models for the Next Generation of Software. <http://oreilly.com/web2/archive/what-is-web-20.html>.

Rowley-Jolivet, Elizabeth / Carter-Thomas, Shirley 2005. Scientific Conference Englishes: Epistemic and Language Community Variations. In Cortese, Giuseppina / Duszak, Anna (eds) *Identity, Community, Discourse. English in Intercultural Settings*. Bern: Peter Lang, 295-320.

Schiltz, Michael / Truyen, Frederick / Coppens, Hans 2007. Cutting the Trees of Knowledge: Social Software, Information Architecture and their Epistemic Consequences. *Thesis Eleven* 89/1, 94-114.

Schmidt, Jan 2007. Blogging Practices: An Analytical Framework. *Journal of Computer-Mediated Communication*. 12/4,1409-1427. <http://jcmc.indiana.edu/vol12/issue4/schmidt.html>.

Sokół, Małgorzata 2012. *Discoursal Construction of Academic Identity in Cyberspace. The Example of an E-Seminar*. Newcastle: Cambridge Scholars Publishing.

Sonvilla-Weiss, Stefan (ed.) 2010. *Mashup Cultures*. Wien, New York: Springer.

van Dijk, Teun A. 2008. *Discourse and Context. A Sociocognitive Approach*. Cambridge: Cambridge University Press.
Vassileva, Irena 2000. *Who Is the Author? A Contrastive Analysis of Authorial Presence in English, German, French, Russian and Bulgarian Academic Discourse*. Sankt Augustin: Asgard.
Wheeler, Steve 2010. Open Content, Open Learning 2.0: Using Wikis and Blogs in Higher Education. In Ehlers, Ulf-Daniel / Schneckenberg, Dirk (eds) *Changing Cultures in Higher Education*. Berlin: Springer, 103-113.
Zimmer, Michael 2009. *Renvois* of the Past, Present and Future: Hyperlinks and the Structuring of Knowledge from the *Encyclopédie* to Web 2.0. *New Media & Society* 11/1-2, 95-113.

Appendix

The corpus used in the study consists of the entries retrieved from eight single-authored academic weblogs available at the following addresses:
(1) <http://the-brooks-blog.blogspot.com/>.
(2) <http://laser.fontmonkey.com/>.
(3) <http://knowledgeandexperience.blogspot.com/>.
(4) <http://walkinthewords.blogspot.com/>.
(5) <http://www.umsu.de/wo/>.
(6) <http://illinois.edu/db/view/25>.
(7) <http://jilltxt.net/>.
(8) <http://phonetic-blog.blogspot.com/>.
When quoted, the authors of the weblogs under investigation are referred to as B1, B2, B3, B4, B5, B5, B7 and B8 respectively.

GIORGIA RIBONI

Twittering Away: *Whole Foods Market* and Conversational Marketing in 140 Characters*

1. Conversational marketing and corporate microblogging

In their 'Cluetrain Manifesto', a document which is widely considered a milestone of marketing studies, Levine *et al.* (2000) urge companies to overcome the credibility crisis in which they find themselves by adopting a more interpersonal communicative style. Starting from the assumption that 'markets are conversations', the authors (or 'cluetrainers') maintain that more personal and interactive ways of getting in touch with potential customers are needed because old corporate traditional communicative practices – prevalently monologic and perceived as rather clearly manipulative – no longer seem very efficient.

Although some of the statements contained in the 'Cluetrain Manifesto' may appear somewhat hyperbolic (Puschmann 2010: 86), it is nonetheless undeniable that over the last decade more and more corporations have embraced a communicative style which encourages participation and dialogue. The strategies which companies adopt to make a virtual connection with potential customers are commonly called 'conversational marketing' (cf. Kelley 2009; De Felice 2010).

Many companies have realized the potential of digital tools and of webgenres, which can be hybridized with traditional selling genres (such as, for instance, the sales letter or the commercial advertisement) for conversational marketing aims. This new approach to promotion appears particularly efficient for a variety of reasons. First of all because Internet communication means are cheaper than others and

* This chapter contributes to a national research project on "Tension and Change in English Domain-Specific Genres" supported by the Italian Ministry of Education (COFIN grant No. 2007JCY9Y9).

can potentially reach an immense audience. Secondly, because virtual conversations about products are carried out in the electronic environment, that is to say in the same environment in which many purchases take place (and possibly in which most, if not all purchases will take place in the future). Moreover, conversational marketing enables corporations not only to advertise their products and services but also provides customers ('users' in the electronic environment) with better quality information and easier access to it. Finally, consumers are no longer the passive target of marketing strategies, but can possibly become a marketing force themselves (De Felice 2010: 53). As a matter of fact, establishing a relationship with customers does not solely mean exchanging opinions and receiving feedback from them, but it also potentially leads them to express their satisfaction for a company or a product on their website and in their virtual network. Word of mouth has always been a decisive factor in orienting customers' buying decisions (Jansen *et al.* 2009: 2169) but after the adoption of Internet tools companies have started utilizing electronic word of mouth (or 'viral marketing')[1] as one of their main conversational marketing strategies (see also Phelps *et al.* 2004).

Among the various rhetorical possibilities offered by the electronic environment, many multinationals have recently discovered and opted for social media and microblogs, as this kind of platform well suits the need to reveal their human side, welcome transparency and forge new relationships with their customers (Gordhamer 2009). Microblogs are short message services of texts, pictures (usually limited to one per message), audio (usually limited to one fragment) and/or video (usually limited to one per message). For the time being, the social communication service Twitter is the leading microblogging platform (Barnes/Böhringer 2009); as a consequence, a significant number of multinationals are exploiting it as a marketing tool (cf. Carraro 2010; De Felice 2010). Launched in July 2006, Twitter is a social network that allows users to "send updates (tweets of 140 characters) to a network of associates (followers) from a variety of devices" (Jansen *et al.* 2009: 2172). The central feature of Twitter,

1 This expression accentuates how rapidly and 'contagiously' the advertising message can be spread (cf. Rayport 1996).

which users see when they log in, is a stream of tweets presented in reverse chronological order (Fig. 1).

Figure 1. Example of Twitter Timeline.

Twitter profiles contain three categories of entries: tweets (or status updates), replies and retweets. Tweets are entries addressed to all followers and to all the visitors of the Twitter page, whereas replies are only addressed to a particular user. Retweets are different from the other two categories because they are not authored, but simply copied and broadcast by microbloggers who like them and want to show their appreciation by further disseminating them. Both retweets and replies are easy to recognize: the former feature the avatar of the original author,[2] whereas the latter typically start with the sign @ followed by the name of the user one intends to communicate with (see examples of retweet and reply in Fig. 2).

2 Before Twitter introduced a 'retweet function' (which offers links under each entry with an automatic retweet option) in 2009, twitterers conventionally signaled that an entry was a retweet by putting the letters RT followed by the sign @ and the name of the original author in front of the message. At present some microbloggers still prefer to put RT@username in front of their retweets because this allows them to display their own avatar and picture.

Figure 2. An example of retweet, reply and tweet.

As Figures 1 and 2 suggest, microblogs are characterized by reverse chronology, frequent updating, and a combination of links with some sort of commentary just like blogs from which they developed (cf., among others, Blood: 2000/2002; Herring *et al.* 2005; Miller/Shepherd 2004/2009; Rosenberg 2009; Myers 2010).

2. Aim, material and method

Taking recent web genre research as a starting point (cf. Askehave/Ellerup Nielsen 2005; Garzone *et al.* 2007; Giltrow/Stein 2009; Miller/Shepherd 2009), this study mainly draws on the methodological contribution of discourse analysis (cf., among others, Brown/Yule 1983; van Dijk 1985; Blommaert 2005) in order to explore the ways in which companies exploit microblogging both rhetorically and linguistically so that it can be turned into an efficient conversational marketing tool. In particular, this research focuses on the participation framework in which corporations address their customers/readers and persuade them to purchase their products. Goffman's (1981) notions of 'animator', 'author' and 'principal' will be applied in order to describe the roles performed by corporate microbloggers in the interaction with their readers.[3]

The Twitter profile of *Whole Foods Market,* an organic food multinational, has been selected as a case study because this Texas-based corporation represents an excellent example of business seriously investing in social media. Its Twitter profile enjoys a very high ranking in the Web.[4] An *ad hoc* corpus, containing all the posts and microblog entries published online on the Twitter page during the period March 10th - May 10th 2010 and consisting of 1,085 entries (totalling 20,317 tokens), has been collected for this research. 88.5% of these entries are replies, 8.5% tweets and 3% retweets.

In order to process this material a top-down approach has been adopted. As a matter of fact, the first part of this chapter investigates the corpus from a predominantly rhetorical point of view (Section 3) and explores the participation framework as well as the communicative functions realized by all the typologies of the entries. The second part of the analysis (Section 4) is instead characterized by a narrower scope and is carried out from a more specifically linguistic perspective: this section is in fact devoted to examining pronominal variation in relation to the different typologies of entries. Since personal pronouns (as the realization of person deixis) encode the relation between sender and receiver, between company and customers, they are an excellent object of study for research on participation frameworks and corporate communication on the Web.

3 Goffman distinguishes the participants of a communicative act according to their role: the 'animator' is the person who physically utters the words, the 'author' is he or she who "selects the sentiments that are being expressed and the words in which they are encoded" and 'principal' is "the individual (or institution) whose viewpoint or position and beliefs are expressed" (Goffman 1981: 144).

4 Twitaholic, a website specialized in assessing the popularity of Twitter profiles according to the number of their followers, estimates that *Whole Foods Market* ranks within the first most followed 100 (data collected during the month of March 2011).

3. Participation framework and rhetorical categorization of the *Whole Foods Market* entries

As Twitter entries are used to realize different rhetorical functions according to the category they belong to (i.e. tweet, reply or retweet), it was decided to treat them separately, devoting a section to each of them.

3.1. Tweets

Instead of describing what is happening to them as most Twitter users do, *Whole Foods Market* microbloggers post tweets aimed at starting a conversation with their readers/customers, building a trusting relationship with them and promoting their brand and its products. This is why it can be affirmed that they not only perform the Goffmanian role of 'authors' because they formulate and post their entries, but they are also 'animators' because they act as mediators between the multinational ('the principal') and the visitors of its Twitter page: unlike most users whom Naaman *et al.* (2010) would term 'meformers' as they primarily focus on themselves and their lives, *Whole Foods Market* microbloggers can be defined 'informers' because their individual identity tends to disappear in their tweets as their main activity is sharing information/promoting their company.

However, the informative purpose of these entries is hardly ever divorced from a typically interpersonal discursive approach, characterized by the continuous attempt to start a contact. Since these entries can potentially reach anybody who surfs the Net and there is no way of knowing who actually accesses the profile, *Whole Foods Market* twitterers rhetorically construct an imaginary readership and address their promotional tweets to it. In the electronic environment the boundaries between the Goffmanian categories of 'ratified participant' (official addressee), 'bystander' and 'eavesdropper' (non-official addressee who casually or secretly listens to the message) are somewhat blurred (Garzone 2007: 20; Myers 2010: 78). Marwick and Boyd

(2010: 129) have coined the term 'networked audience' to refer to the stratified and interconnected audience of Twitter entries:

> The networked audience [...] consists of real and potential viewers for digital content that exist within a larger social graph. These viewers are connected not only to the user, but to each other, creating an active, communicative network. [...] In contrast to the imagined broadcast audience, which consumes created content with limited possibilities for feedback, the networked audience has a clear way to communicate with the speaker through the network. [...] Just as writers fictionalize the audience within the text in their audience addressed, Twitter users speak directly to their imagined audience .[...] They target tweets to specific audience members, and conceal or reveal information based on who they imagine to be listening.

The strategies deployed by *Whole Foods Market* microbloggers to address this imaginary audience include the use of features which typically occur in conversation, like for example the high frequency of deictic constructions (especially person deixis; cf. Section 4), the presence of questions (rhetorical or otherwise) and imperative forms. As Myers underlines, questions and imperatives allow the text producer to make addressees feel involved in the communicative process because "any utterance that isn't a statement assumes there is someone else there to respond; a question calls for an answer, and a directive calls for an action" (Myers 2010: 82). These kinds of rhetorical devices are used to activate the phatic function in all the typologies of entries posted on the *Whole Foods Market* profile. As the following examples demonstrate, all thematic categories of tweets (the explanation of the properties of a certain fruit or plant, the healthy tip, the recipe and the celebration of a festivity) contain a more or less overt informative/promotional component as well as a question (example 1), an imperative (examples 2-4-5) or some typically conversational features ("Thanks" in example 3)[5].

(1) Strawberries are a truly versatile springtime treat. *What's your favorite way to eat them?* (explanation of the beneficial properties of a plant whose extract is sold at Whole Foods Market stores).

---

5 Courtesy of *Whole Foods Market. 'Whole Foods Market'* is a registered trademark of *Whole Foods Market IP, L.P.*

(2) *Be Good to Your Whole Body* with healthy tips for women from Dr. Tori Hudson. http://cot.ag/9Yo4RB (healthy tips).

(3) Happy Mother's Day! For all the moms out there – *you rule! Thanks for all that you do*! http://cot.ag/92uury (celebration of a festivity).

(4) Some (not all) of our stores are even offering tax-free shopping today! *Check w/ your local store for more details*. http://tr.im/allstore (company's announcement).

(5) *Try a healthier twist on an old favorite* with our recipe for Carrot-Oat Cake! http://cot.ag/c68OiO (recipe) (emphasis added).

The adoption of a *you*-style characterized by the frequent use of questions, answers and imperatives is not a distinctive trait of corporate microblogging, but is typical of promotional discourse. In particular, it is possible to find interpersonal discursive features in most traditional selling genres like the sales letter, the sales talk, the commercial advertisement etc. As a consequence, the presence of such an interactional style on the *Whole Foods Market* profile is hardly surprising as it is probably due to the existence of an interdiscursive relation between corporate twittering and promotional genres.

3.2. Replies

The rhetorical construction of the audience represents the most significant difference between tweets and replies: whereas the former are addressed to an imaginary readership, the latter are sent to a specific user. Replies are easy to identify because they are preceded by the sign @ followed by the name of the user one intends to talk to. On the Twitter platform, the @user form is provided with a link to the intended recipient of the message, so that he/she is immediately informed about it and the sender can gain his/her attention (Honeycutt/ Herring 2009). In the case of replies, it is possible to distinguish between 'ratified' and 'unratified' participants of the conversation, because microbloggers explicitly mention the recipient of their message.

Just like tweets, most of the replies posted on the *Whole Foods Market* Twitter account contain links to the blog (or the company's website). Most replies are about customers' problems, requests and enquiries. Since tweets addressed to a specific user are by far the most common type of entry, then it can reasonably be concluded that *Whole Foods Market* mainly uses its microblog for customer care purposes, as the following examples demonstrate:

(6) @IGREJAB Most of our stores do carry polenta. Your best bet is to call first and ask just to be sure – http://tr.im/allstore

(7) @studes You can check in with @WholeFoodsATX for nutritional info for the Big John cookies.

(8) @mayasuri We generally do not have positions open for anthropologists, but are not against hiring them for other positions.

Unlike answering requests via email or by phone – that is to say by dealing with customers' request privately and individually using Twitter for customer care purposes enables *Whole Foods Market* to show all possible webpage viewers the company's serious concern with customers' satisfaction. Moreover, the fact that replies, unlike blog comments, can be directly accessible to anyone, makes this kind of interactional exchange similar to the genre of Q&A. The innovation introduced by microblogs lies in the fact that the interaction not only provides 'push communication' (i.e. information obtained after having contacted someone), but it also enables 'pull communication' (that is to say knowledge derived from reading somebody's postings) (Günther *et al.* 2009: 1; De Felice 2010: 23). Differently from what happens with other types of web platforms (for instance with blogs), Twitter allows real-time and many-to-many (networked) communication: for this reason early adopting companies like *Whole Foods Market* have realized and exploited its potential as a way of providing customer care (cf. De Felice 2010).

3.3. Retweets

Since retweeting on Twitter is relatively new, the conventions for this practice are not well established yet. However, by copying and diffusing a text outside its original rhetorical situation, twitterers alter its initial communicative purpose and intended audience. As a matter of fact, those microbloggers whose tweets are picked up and retweeted by *Whole Foods Market* are typically 'meformers' (Naaman *et al.* 2010) who describe what they are doing or have done. Examples 9 and 10 show that the user's experience plays a crucial role in this kind of tweets: in this respect, the recurrence of the first person pronoun occupying the subject position is one of their typical features[6]:

(9) RT @Salsalover *I* absolutely love how every employee at @WholeFoods is nutritionally informed – had a 15 min conversation with the baker about sprouted bread.

(10) RT @Agnes @Lilith *I* <3 @WholeFoods. It's the only place *I* go where they don't look at me like *I*'m crazy when *I* ask for vegetarian things (emphasis added).[7]

A possible reason why *Whole Foods Market* microbloggers retweet other users' entries has to do with the company's necessity of finding a more personalized way of communicating with its consumers: tweets are picked up and retweeted, on the one hand, to express the individual point of view of the user who has conceived them, and, on the other hand, to add a more 'human note' to the otherwise fairly impersonal *Whole Foods Market* Twitter profile. If Goffmanian categories are applied, it can be affirmed that company microbloggers drop the role of 'author' and only act as animators as they retweet somebody else's entries.

As examples 9 and 10 show, the tweets selected for retweet typically contain a mention of the multinational (usually realized with the '@WholeFoods' form). This seems to suggest that one possible reason why *Whole Foods Market* retweets its customers' entries is to show that it is interested in what they say about it. Example 10 is a retweet of what the author of the entry (Agnes) replies to one of her Twitter contacts (Lilith); *Whole Foods Market* was not the intended recipient of this message, but since it was cited in it, the company's microbloggers were notified (and decided to spread amanda_collier's

6 Person pronouns are discussed in the following section.
7 Unless otherwise indicated, all websites were last accessed on 3 June 2011.

tweet among their own audience). Microblogging platforms offer the possibility of 'quietly listening' to what customers say about brands so that the latter can respond to it. Consultants use the term 'listening marketing' (Carraro 2010: 46) to define this promotional strategy.

If a tweet is remarkably positive about *Whole Foods Market,* then the company's twitterers retweet it for all their followers to read. In this way, not only can they prove that *Whole Foods Market* is attentive to what its consumers think about it, but they also amplify and spread tweets to a new (and possibly larger) audience; as Boyd *et al.* (2010) point out:

> Retweets can knit together tweets and provide a valuable conversational infrastructure. Whether participants are actively commenting or simply acknowledging that they're listening, they're placing themselves inside a conversation. Even when they are simply trying to spread a tweet to a broader audience, they are bringing people into a conversation.

Apart from using retweets as a way to manifest interest in what users think, *Whole Foods Market* exploits this kind of entry as electronic word of mouth (Jansen *et al.* 2009). As a matter of fact, by diffusing somebody else's favourable impressions to its followers, the company can show to its Twitter visitors that customers are satisfied with its products. As discussed in Section 1, electronic word of mouth is extremely cost-effective, and can therefore be considered a successful strategy.

Whereas the reasons for *Whole Foods Market* to publish other users' flattering tweets on its profile are self-evident, consumers' motives for tweeting something enthusiastic about the company may appear more obscure. In actual fact, people may decide to write something positive about *Whole Foods Market* in the hope that it will be retweeted so that their username and avatar will appear on one of the most famous microblogging profiles (it has almost two million followers) and will receive wider public exposure. Just as *Whole Foods Market* seems to exploit Twitter's communicative conventions for promotional purposes, those users who post flattering comments on the company make use of this microblogging platform to express their own 'private intentions' (cf. Bhatia 1993; 1995; 1997; 2004).

4. Use and function of personal pronouns in the *Whole Foods Market* microblog conversations

As already pointed out, the *Whole Foods Market* microblog is used for conversational marketing purposes and aims to engage its readers in virtual conversation. Consequently, the high frequency of first and second person pronouns in the Twitter page is hardly surprising: as a matter of fact the *Whole Foods Market* staff tend to establish a tri-part relationship between themselves, their company as a whole and their readers. However, the uses and occurrence of *I*, *we* and *you* vary depending on whether they are contained in a tweet, a reply or a retweet. Table 1 summarizes the referents of the first and second person pronouns. The column corresponding to retweets is almost empty, as this category includes entries written by many different authors for many different purposes; consequently, it was impossible to identify a common referent for the pronouns *you* and *we*.

	Tweets	*Replies*	*Retweets*
You			
	generic single reader		
	general audience		
		specific reader @user	
We			
	company, microblogger and reader (inclusive *we*)		
	company and microblogger (exclusive *we*)	company and micro-blogger (exclusive *we*)	
I			
	/	author (microblogger) (rare)	animator (author of the tweet)

Table1. Referents of the first and second person pronoun.

4.1. You

In tweets and replies, *you* is the most frequently occurring pronoun as a possible indication of the fact that corporate microblogs are primarily devised to establish contact with customers or potential customers. Since the *Whole Foods Market* microbloggers seem aware of the fact that they are dealing with a networked audience, they vary their use of the second person pronoun according to the communicative purpose they want to achieve. In order to establish contact and promote their products in tweets they utilize *you* to refer to both a generic single reader (thus making him/her feel involved) and the entire audience:

(11) Happy Cinco de Mayo! Felicidades! How will *you* be celebrating?

(12) Did *you* know that natural cork oak forests support one of the world's highest levels of forest biodiversity? (Emphasis added).

As the above examples show, the second person pronoun is often to be found in questions which have a phatic function as they are meant to elicit some kind of response or feedback from the audience, thus giving rise to an exchange of comments or replies[8]. When authors succeed in starting a conversation, they utilize the second person pronoun to address a specific reader, as happens in replies:

(13) @DietStarts2mrw See *you* soon at @WholeFoodsCHI! No need to blow *your* whole paycheck if *you* check out http://tr.im/wholedeal.

(14) @everywhereist If *you* let Customer Service at the store know next time *you*'re in, we can do something to make it up to *you*. Thanks! (emphasis added).

The percentage frequency of occurrence of *you* in the replies (1.80%) almost doubles the percentage frequency of *you* in the tweets (0.96%). Such a difference in figures seems to suggest that only replies repre-

---

8 Since Twitter entries are very similar to conversational exchange, their formulation is sometimes careless, as demonstrated by the following examples of tweets containing an ellipsis of the subject and the operator:

- Think cutting pineapples is a task better left to a chef? Have a hard time shopping for a good one? […]
- Want to win a five CD set from Putumayo's Latin collection?

sent the genuinely 'conversational' Twitter type of entry, whereas tweets and retweets are much more promotional in nature.

4.2. We

The pronoun *we* can be used either exclusively, (i.e. to indicate the text producer and the people belonging to his/her group, but excluding the addressee/s) or inclusively, (i.e. to indicate the texts producer, the people belonging to his/her group and the addressee/s; cf. Maitland/ Wilson 1987; Dieltjens/Heynderickx 2007). The analysis of the *Whole Foods Market* Twitter entries highlights that inclusive *we* seldom occurs in the tweets. However, it is important to note that the imperative form *let's* is among the most frequently recurring expressions both in the posts and in the tweets: the personal pronoun *us* is almost always used inclusively, to suggest that the company and its consumers share common objectives and interests, as the following examples demonstrate:

(15) *Let's choose* growers who take the road less traveled [...].

(16) *Let's cook* in the kitchen, not the science lab [...]. (Emphasis added)

However, the *Whole Foods Market* microbloggers tend to use the first person pronoun almost always to refer only to themselves and their company and not to their readers; in this way they can downplay the distance between themselves as 'authors' and 'animators' and their 'principal', *Whole Foods Market.* In their tweets the multinational appears as a collective *we*, a single entity which interacts with its customers, asks for or offers advice, spreads pieces of news, details future plans and promotes its products:

(17) *We*'ve got a *shocking* new mobile app that'll help *you* keep *your* diet on track [...].

(18) *We*'ll show *you our* standards (http://tr.im/wfmqs) if *you*'ll show *us yours.* What are they?

(19) *We* never use artificial sweeteners, but *we* have sweet alternatives. What are *your* favorites? (Emphasis added)

The frequent occurrence of the personal pronouns and the possessive adjectives/pronouns in the examples above highlights the bloggers' desire to involve their audience (which is repeatedly addressed) and to emphasize the relationship existing between the latter (identified by the pronoun *you*) and the company (*we*).

4.3. I

In the *Whole Foods Market* microblog the pronoun *I* occurs considerably less than its plural counterpart. As regards tweets, the first person singular is totally absent, possibly because microbloggers only want to communicate on behalf of their company and tend to stress that they are not the 'principal' of the communicative act. The fact that *Whole Foods Market* twitterers never refer to themselves in their tweets is quite meaningful, especially in the light of the great importance microblogs typically attribute to the identity of the author (cf. Miller/Shepherd 2009: 268; De Felice 2010: 38).

However, this pronoun is fairly common in retweets and it is possible to find some instances of *I* in replies which represent virtual conversational exchanges between customers and microbloggers.[9] This indicates that in these exchanges which are used by the company to deal with customer care issues online, the *Whole Foods Market* twitterers prevalently speak on behalf of their company, but they also sometimes take personal responsibility in helping consumers:

(20) @LevanaCooks: Sure, *I'*ll let our Exclusive Brands team know about the hard to read label and *we* 'll see what *we* can do!

(21) @JPetersonPhotog: Which local store do you shop? *I* can ping *them* about their Prepared Foods labeling.

In both examples 20 and 21 the microbloggers use the first pronoun singular to stress their own individual contribution in assisting customers, possibly with the aim to emphasize how well-organized and

9 In the 960 replies collected for this study it was possible to find 20 instances of *I*, whereas in the 32 retweets of the corpus this pronoun occurs 15 times.

interconnected their company is. Secondly, by underlining their role as mediators between the public and the *Whole Foods Market* branch that will practically deal with requests or complaints, they call attention to the rapidity and efficiency with which the corporation tackles its customers' problems.

In retweets the first person pronoun is rather frequent, as this kind of entry is produced by other twitterers who use microblogging to talk about themselves, their life and experiences (cf. Section 3). As a result, the interpretation of the first person pronoun outside the original context of the retweet (i.e. on the profile of the user who has posted it) may appear somewhat complicated: does *I* refer to the person who has published the entry or to the *Whole Foods Market* microblogger who has copied and pasted it on the company's Twitter page to suggest that he/she (therefore indirectly the company he/she represents) thinks likewise?

As Boyd *et al.* (2010*)* point out, such an ambiguity does normally exist. However, on the *Whole Foods Market* most retweets are posted by users who do not work for the company[10]. As a consequence the referent of the first person singular hardly ever coincides with the company itself or its employees, as shown in the following examples:

(22) RT @gilad: Check out some of my goodies *I* picked up this morning at @wholefoods.

(23) RT @barbiegirl: *I* cannot stop eating @WholeFoods Organic Peanut Butter Pows cereal. (Emphasis added)

If the *I* of the person who posts the tweet and the *I* of the company's microblogger (i.e the 'author' and the 'animator" of the message) refer to different users, then these entries are supposedly written by someone outside the corporation (the principal), and are therefore theoreti-

10 Due to the difficulties of establishing authorship in the electronic environment, it was impossible for me to verify whether the retweets posted on the *Whole Foods Market* Twitter page had been written by somebody paid by the company to do so or whether they were authentic. However, the significant number of retweets posted on the *Whole Foods Market* profile as well as the fact that they were originally published on a variety of different accounts legitimately induce to think that at least some of these entries are genuine.

cally disinterested and unbiased. That is why retweets are exploited by *Whole Foods Market* as a form of electronic word of mouth.

5. Concluding remarks

In this chapter the Twitter profile of *Whole Foods Market* has been analysed in order to explore how the company exploits social media to establish contact with its consumers. The analysis has highlighted the fact that, in order to communicate and market their products efficiently, the company's microbloggers tend to strategically change the participation framework in which they operate.

Since it is impossible to distinguish between ratified and unratified participants, when writing entries *Whole Foods Market* twitterers rhetorically construct an imaginary audience which they address and try to involve by utilizing the language typically occurring in conversations. Their promotional tweets are formulated in the first person pronoun *we* (indicating the contiguity between the author/animator and the principal of the message) and directed at the receiver, a generic *you.*

In replies, instead, the recipients of the message are explicitly mentioned so that it is possible to identify them as the ratified addressees. Whereas tweets are used to start a virtual conversation with the readership and are therefore used as a promotional tool, replies are mostly utilized as a way of providing customer care.

Finally, retweets are different from all the other entries because they are not 'authored' by the *Whole Foods Market* microbloggers themselves, but by somebody else. This is why retweets probably represent the type of entry which can be best turned into an electronic-word-of-mouth tool. As discussed in Section 4.3., the fact that the principal, the animator and the author of the message are three distinctive figures makes the promotional content of the retweet all the more persuasive.

In conclusion, the skilful use of various participation frameworks and of person deixis contributes to making Twitter a valid conversational marketing tool. However, the adoption of a *you*-style which aims to promote products and services by starting an interaction with potential buyers does not solely belong to the conventions of corporate microblogging and conversational marketing, but derives from traditional promotional genres such as the sales letter, sales talk, the commercial advertisement etc. The presence of a *you*-style is therefore probably due to interdiscursivity with these genres which, just like corporate microblogging, aim to advertise a company's products by engaging its customers in conversation.

On the other hand, this study has also highlighted that those characteristics which are specific to microblogging (for instance the fact that it allows real-time conversations) have induced a multinational like *Whole Foods Market* to hybridize the corporate blogging genre with less overtly promotional genres such as customer care and Q&A.

These observations contribute to making the generic status of microblogging highly uncertain as well as rather difficult to establish. Further research into this subject may be helpful to shed some light on the evolutionary stage in which blogging and microblogging find themselves and to determine whether these two types of platform still belong to the same genre or whether this is no longer the case.

References

Askehave, Inger / Ellerup Nielsen, Anne 2005. Digital Genres: A Challenge to Traditional Genre Theory. *Information Technology & People.* 18/2.

Barnes, Stuart J. / Böhringer, Martin 2009. Continuance Usage Intention in Microblogging Services: The Case of Twitter. *Proceedings of the 17th European Conference on Information Systems (ECIS).* Available at <http://www.ecis2009.it/papers/ecis2009-0164.pdf>.

Bhatia, Vijay, K. 1993. *Analysing Genre: Language Use in Professional Settings.* London: Longman.

Bhatia, Vijay K. 1995. Genre-mixing in professional communication: the case of 'private intentions' v. 'socially recognized purposes'. In Bruthiaux, Paul / Boswood, Tim / Du-Babcock, Bertha (eds) *Explorations in English for Professional Communication.* Hong Kong: City University of Hong Kong, 1-19.

Bhatia, Vijay K. 1997. Genre-Mixing in Academic Introductions. *English for Specific Purposes*: 16/3, 181-196.

Bhatia, Vijay K. 2004. *Worlds of Written Discourse.* London/New York: Continuum.

Blommaert, Jan 2005. *Discourse: A Critical Introduction*, Cambridge: Cambridge University Press.

Blood, Rebecca 2000. *Weblogs: A History and Perspective,* 2000. Available at <http://www.rebeccablood.net./essays/weblog_history.html>.

Blood, Rebecca 2002. *The Weblog Handbook. Practical Advice on Creating and Maintaining your Blog,* Cambridge, MA: Perseus Publishing.

Boyd, Danah / Golder, Scott / Lotan, Gilad 2010. Tweet Tweet Retweet: Conversational Aspects of Retweeting on Twitter. *Proceedings of the 43th Annual Hawaii International Conference on System Sciences* HICSS-43, IEEE: Kauai, HI. Available at <http://www.danah.org/papers/TweetTweetRetweet.pdf>.

Brown, Gillian / Yule, George 1983. *Discourse Analysis.* Cambridge: Cambridge University Press.

Carraro, Massimo 2010. *Un etto di marketing. È un etto e mezzo, lascio?* Milano: Alpha Test.

De Felice, Luca 2010. *Marketing Conversazionale. Dialogare con i clienti attraverso i social media e il Real-Time Web di Twitter, FriendFeed e Facebook.* Milano: Il Sole 24 Ore.

Dieltjens, Sylvain / Heynderickx, Priscilla 2007. Strategic Uses of the Pronoun We in Business Communication. In Garzone, Giuliana / Sarangi, Srikant (eds) *Discourse, Ideology and Specialized Communication*. Bern: Peter Lang, 233-247.

Garzone, Giuliana / Poncini, Gina / Catenaccio, Paola (eds) 2007. *Multimodality in Corporate Communication. Web Genres and Discursive Identity.* Milano: FrancoAngeli.

Garzone, Giuliana 2007. Genres, Multimodality and the World Wide Web: Theoretical Issues. In Garzone/Poncini/Catenaccio (eds), 15-30.

Giltrow, Janet / Stein, Dieter (eds) 2009. *Genres in the Internet.* Amsterdam/Philadelphia: John Benjamins.

Gordhamer, Soren 2009. 4 Ways Social Media is Changing Business. *Mashable* Available at <http://mashable.com/2009/09/22/social-media-business/>.

Goffman, Erving 1981. *Forms Of Talk.* Philadelphia: University of Pennsylvania Press.

Günther, Oliver / Krasnova, Hanna / Riehle, Dirk / Schöndienst, Valentin 2009. Modeling Microblogging Adoption in the Enterprise. *Proceedings of the 15th Americas Conference on Information Systems.* Available at <http://warhol.wiwi.hu-berlin.de/aigaion2/index.php/publications/show/227>.

Herring, Susan C. / Scheidt, Lois Ann / Wright, Elijah / Bonus, Sabrina 2005. Weblogs as a bridging genre. *Information Technology & People* 18/2, 142-171.

Honeycutt, Courtenay / Herring, Susan C. 2009. Beyond Microblogging: Conversation and Collaboration via Twitter. *Proceedings of the Forty-Second Hawaii International Conference on System Sciences* (HICSS-42). Los Alamitos, CA: IEEE Press. Available at <http://ella.slis.indiana.edu/~herring/honeycutt.hering.2009.pdf>.

Jansen, Bernard J. / Zhang, Mimi / Sobel, Kate / Chowdury, Abdur 2009. Twitter Power: Tweets As Electronic Word Of Mouth. *Journal Of The American Society For Information Science And Technology.* 60/11, 2169-2188.

Kelley, Braden 2009. Conversational Marketing White Paper. *Blogging Innovation,* Available at <http://www.business-strategy-innovation.com/wordpress/>.

Levine, Rick / Locke, Christopher / Searls, Doc / Weinberg, David 2000. *The Cluetrain Manifesto: The End of Business as Usual.* Available at <http://tinyurl.com/3ym8je>.

Maitland, Karen / Wilson, John 1987. Ideological Conflict and Pronominal Resolution. *Journal of Pragmatics.* 11, 495-512.

Marwick, Alice, E. / Boyd, Danah 2010. I tweet honestly, I tweet passionately: Twitter users, context collapse, and the imagined audience. *New Media & Society*, February 2011; 13/1, 114-133.

Miller, Carolyn R. / Shepherd, Dawn 2004. Blogging as a Social Action. *Into the Blogosphere.*2004. Available at <http://blog.lib.umn.edu/blogosphere/blogging_as_social_action.html>.

Miller, Carolyn R. / Shepherd, Dawn 2009. Questions for Genre Theory From the Blogosphere. In Giltrow, Janet /Stein, Dieter (eds) *Genres in the Internet.* Amsterdam/Philadelphia: John Benjamins, 263-290.

Myers, Greg 2010. *Discourse of Blogs and Wikis.* London/New York: Continuum.

Naaman, Mor / Boase, Jeffrey / Lai, Chih-Hui Lai 2010. Is It Really About Me? Message Content in Social Awareness Streams. *Proceedings of CSCW 2010.* Available at <http://infolab.stanford.edu/~mor/research/naamanCSCW10.pdf>.

Phelps, Joseph E. / Lewis, Regina / Mobilio, Lynne / Perry, David / Raman, Niranjan 2004. Viral Marketing or Electronic Word-of-Mouth Advertising: Examining Consumer Responses and Motivations to Pass Along Email. *Journal of Advertising Research.* 44, 333-348.

Puschmann, Cornelius 2010. *The Corporate Blog As An Emerging Genre of Computer-Mediated Communication: Features, Constraints, Discourse Situation,* Göttingen: Universitätsverlag Göttingen.

Rayport, Jeffrey 1996. The Virus of Marketing. *Fast Company,* Available at <http://www.fastcompany.com/magazine/06/virus.html>.

Rosenberg, Scott 2009. *Say Everything. How Blogging Began, What's Becoming and Why It Matters.* New York: Crown Publishers.

van Dijk, Teun A. (ed.) 1985. *Discourse and Communication,* Berlin/New York: de Gruyter.

Websites

Twitaholic <http://twitaholic.com/>.
Webopedia <http://www.webopedia.com/>.
Whole Story Blog <http://blog.wholefoodsmarket.com>.
Whole Foods Twitter Profile <http://twitter.com/wholefoods>.

MARIA CRISTINA PAGANONI

Online Branding from Hybrid Ads to Corporate Twittering

1. Study design: branding on Twitter

Among the latest trends in consumer culture, a most notable one is the incorporation of advertising into the wider framework of branding (Fabris 2008), which can be summarized as an all-encompassing set of promotional discourses and practices that, going much beyond mere persuasion, associates a tangible or intangible good with specific values and lifestyle patterns. This trend has doubtless been amplified by the migration of advertising genres from old to new media and their concomitant transformation brought about by the enhanced multimodality and interactivity of Web 2.0. Both features have boosted the role played by consumers' feedback in highlighting how a brand's immaterial value may be related to processes of identity construction mediated by consumption patterns. Within the highly mobile Internet environment, social networks have further contributed to the dissemination of branding onto multiple platforms of conversation.

Ubiquitous branding and the monitoring of user-generated content have radically transformed advertising strategies and campaigns. The communicative phenomenon within which this present analysis is situated is the emergence of 'hybrid ads', i.e. "advertising that combines promotion of products and services while communicating a general message about the company" (Koteyko 2009: 113), due to the fact that persuasion has become "the dominant but definitely not single communicative function of Internet advertising" (Koteyko 2009: 114). In addition to this, the attempt to involve customers in a dialogic relationship with the company itself and favour their active engagement over the canonically 'push' efforts of marketing, i.e. spreading the word out to all potential buyers in order to hit the few interested ones, is clearly facilitated by the 'social networking' turn, brought about by the development and intensification of social media and the novel configurations of online communities of consumers, or rather 'prosumers' (Toffler 1980), that play an active role in content generation. Especially in the case of large companies keen on innovation, platforms for both content creation and discovery – such as blogs, Facebook, Flickr, Twitter and YouTube – integrate social media with corporate websites, fitting social networking into the marketing mix.

The overlap of the informing, educating and entertaining functions and customer feedback should be regarded as a manifestation of an emerging "genre ecology of the Internet" (Herring *et al.* 2004: 1, quoting Erickson 2000). The word 'ecology' for 'system', borrowed from situated genre theory, is here employed to underline the dynamism of web-mediated genres and to propose a descriptive rather than prescriptive analytic approach, flexible enough to accommodate less formalized, but nonetheless powerful formats that would not otherwise fit into conventional taxonomies. Part of this new environment is the fact that "media consumers increasingly participate in the process of designing, producing and marketing media content and experiences" (Banks/Potts 2010: 253), adding their 'playlabour' (Banks/Potts 2010: 260) and thus contributing to the redefinition of promotional corporate genres.

The latest newcomer among social networks, sprouting in 2006 from the minds of three West Coast college dropouts in their early thirties – Jack Dorsey, Biz Stone and Evan Williams – Twitter has taken the web by surprise. Categorized as 'micro-blogging' by partial analogy with longer and more expansive blogging, a tweet limits the number of its characters to the bare figure of 140 or less and adapts that particular form of discourse known as 'texting'[1] to a huge social networking component. The Twitter stream appears in a public time-

---

1 A text message can be longer than a tweet as it allows for 160 characters. The fact that Twitter is based on texting does not limit its scope to that of a 'words-only' medium, as tweets can circulate links to all kinds of multimodal content. It is rather its web interface which is primarily text-based. Apart from contingent text-message fees, Twitter remains a free-of-charge service.

line where tweets, if unprotected, can be read and shared by everyone. Unlike blogging, 'twittering' discards any hierarchical structure due to the presence of a main author (the blogger) proposing selected topics to readers, its public timeline (i.e. the webpage where all tweets in every language are collected) resembling instead a collective 'stream of consciousness', an explosion of thoughts, comments and signals.

In the innovative scenario of co-creative media culture (Jenkins 2006; Israel 2009), the role of Twitter in the evolution of online branding is more and more visible, as an increasing number of companies are now learning to leverage Twitter to help their business grow, for several good marketing reasons. Firstly, a link on Twitter is a free, easy and cheap way to promote a new product, website page, blog post, or other updated content. Secondly, the social network complies well with mobile phone usage, regarded as the new frontier of marketing and sociability.[2] As Kuwabara *et al.* (2002: 3270) observe:

> connectedness oriented communication does not focus on the reproduction of message contents, but on the social relationships, expected to be formed as a result of communication activities. Therefore, in connectedness oriented communication, there is no need to transmit a large amount of data.

Thirdly, thanks to its networking logic, Twitter has the potential to recruit an expanding number of customers, operating in this way as a cheap marketing tool that does not require spending a large portion of the company's budget. And, finally, "as the leading mobile microblogging platform it reveals real-time public reaction and sentiment", working as a 'free focus group' (Bendandi 2009: 22).

This challenging option explains why major brands, such as Coca-Cola, Dell, Microsoft, Starbucks and WholeFoods, to name but a few, have begun to open official corporate accounts on CoTweet. Self-described as "a platform that helps companies reach and engage customers using Twitter" (<http://cotweet.com>),[3] CoTweet turns out to be very suitable to proactive marketing and team working in the

---

2 Besides the powerful tool of viral marketing, mobile connectedness, if integrated with locative media, has the potential for businesses to address customers on the basis of their geographical location.

3 Unless otherwise indicated, all websites were last accessed on 31 October 2011.

management of online conversations between a company's communication team and its customers. In addition to this, "companies are offering to pay Twitterers to gain access to their followers" (Watkins 2009), in other words tapping Twitter users' own contacts to send tweets containing ads. What is emerging is that Twitter is now integrating and interacting with the entirety of persuasive appeals retraceable in advertising video clips, banner ads, forums and blogs.

2. Research questions and methods

Probably owing to fast-paced technological innovation and creative violation of generic conventions, web-mediated advertising is still largely under-researched (Schumann/Thorson 2007), while new media experts warn that social network analysis to measure advertising effectiveness is far from being an easy task (Starling 2009).

Against this extremely fluid background which still needs to be mapped, the purpose of this study is to offer a preliminary analysis of micro-blogging and its linguistic and discursive encoding by concentrating on the emerging promotional genre of corporate tweets on company websites. The research is moved by the curiosity to investigate and understand the textualization processes of this peculiar genre and assess its branding effectiveness against its constraining conciseness on the one hand and the semiotic plenty of multimodal online advertising on the other.

The object of analysis is the Coca-Cola Company website. The most recognized brand in the world, Coca-Cola largely owes its popularity to over one hundred years of successful advertising, which began in 1886 when the drink was first produced and sold in Atlanta. Reflecting its iconic status in mass culture, the company website stands out as a breath-taking archive of post-modern, multi-purpose marketing for its wealth of brilliantly-crafted, state-of-the-art branding resources – from pop culture and brand history quotations to alluring design, video clips and invitations to join social networks. This extremely varied portal introduces what can be credibly named the Coca-Cola universe, containing from heritage to collectibles, from a sustainability section to – of particular interest for our analysis of advertising formats – a corporate blog, the Coca-Cola Conversations, launched on January 23, 2008 and maintained by Philip Mooney, the company's official archivist.

The website (cf. Figure 1) skillfully mixes presentation, information, entertainment and educational designs, as already pointed out by Koteyko (2009) for yoghurt advertising. It contains different genres employed for their persuasive appeal, from press releases to company news and advertising trailers, from the 'conversations blog' to a link to Twitter, characteristically placed between the 'inform' and 'learn' sections.

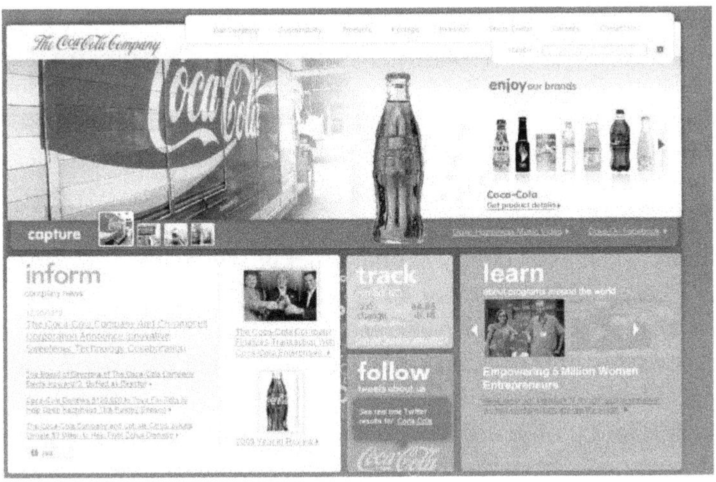

Figure 1. The Coca-Cola Company website.

As the company has set up several Twitter accounts, this study focuses on the one that appears most active in engaging customers – @CocaCola – the "Official tweets of Coke & The Coca-Cola Company", which contains a link to the company website. To the purpose of this study, a sample of over 700 tweets in English (disregarding those in Spanish, the other language regularly used in the messages) was retrieved from April 30 to May 31, 2010 from the corresponding profile timeline (i.e. the micro-blog associated to this specific account) and analyzed in the context of the other promotional material published on the website (especially the blog posts and comments archived on a permanent link of the blog's homepage), constantly keeping in mind the brand's long-standing advertising history. The following research questions were addressed:

- What are the distinctive generic features of corporate tweets as an advertising genre and what kind of textualization processes can be described therein?
- As regards corporate branding, what is the relationship of tweets with the rest of the promotional material published on the Coca-Cola Company website?
- What are the results of this preliminary search on genre hybridization in micro-blogging?

The multidisciplinary methodological framework here espoused combines contributions from studies on computer-mediated communication (Werry 1996; Erickson 2000; Simpson 2005; Boyd/Ellison 2007; Honeycutt/Herring 2009; Page 2010), ICT (Kuwabara *et al.* 2002; Jackson *et al.* 2007; Schroeder 2010) and online advertising (Arvidsson 2006; Schumann/Thorson 2007; Koteyko 2009; Banks/Potts 2010) with those of Genre Analysis (Berkenkotter/Huckin 1995; Bhatia 1993, 2000, 2002, 2004). This framework is enriched in turn with insights derived from Critical Discourse Analysis (Fairclough 2003). Arguably, the main asset of this multifaceted approach lies in the meaningful connections it establishes between text typologies and society, by showing how the transformation of genre functions and of discursive and rhetorical strategies originates from much wider social processes.

Regarding the issue of genre diachronicity and societal change, it should be observed that the rapid evolution of online advertising does not erase antecedent genres in favour of recent ones. Rather, the perceivable phenomenon is that of accretion, i.e. of the mutual coexistence and synergy of different promotional genres across multiple channels out of marketing reasons. The current business philosophy addresses the following issue: "we need to reach everyone in the forum and format they choose – not what we choose" (Bendandi 2009: 23). The observable consequence is a much more ambitious brand construction that blurs the boundaries between consumers and products, the latter increasingly a marker of added value inscribed in the collective imagination and expectations. This discursive and semiotic colonization of the symbolic system explains, among other things, why the ambitious 'Open happiness' philosophy, extolled in the Coca-Cola Company's recent advertising campaign, resonates down to the tiniest corporate tweets as well.

(1) Did you have a Coke and smile today? ^OP 1:56 PM May 12th via CoTweet.

(2) Are you happy? What makes you happy? See the results of our Coca-Cola Global Happiness Study: http://CokeURL.com/4pxv ^OP 8:17 AM May 11th via CoTweet.

3. Data analysis

A flexible enough description of how Twitter works is that offered by Boyd and Ellison (2007) with their definition of social network sites as "web-based services that allow individuals to (1) construct a public or semi-public profile within a bounded system, (2) articulate a list of other users with whom they share a connection, and (3) view and traverse their list of connections and those made by others within the system". To join the conversation on Twitter, users are required to register by inserting their details, creating a username and opening an account which generates a public profile webpage with the address <http://twitter.com/username>. The tweets sent by a user and those received by his/her followers, i.e. other users that have added that account to their contacts, appear listed in reverse chronological order in the timeline embedded within the homepage (see the central column in Figure 2).

Figure 2. The CocaCola profile timeline.

Tweets are based on an easily recognizable template according to their main function, which normally falls under one of the three following categories:

- *status update*. This is in a sense the primary Twitter function, a 'one-to-many' message that answers the fundamental question expected to trigger the conversation, i.e. 'What's happening?' Updates serve the communicative purposes of engaging customers by sharing promotional information about the company as a form of branding.
- *reply* to another tweet, an action comparable to that of answering an SMS message, whose communicative purposes involve maintaining contact with individual customers while promoting the brand with the entire online community of followers.
- *retweet*, i.e. forwarding a tweet to followers on the grounds of its perceived informational and promotional appeal. Retweets are signalled by an appropriate symbol, or by the 'RT' acronym, plus @username. As text types, retweets can be brought back to status updates or replies.

When the Twitterer is an organization or company, status updates amount to brief announcements that release news, information and ideas about anything that may shed a positive light on the company to its network of followers, such as awards, innovations, anniversaries and events in general. Although they do not usually expect a direct response, status updates are quite effective at breaking corporate news to which followers can react in real time, as their brevity makes them much more expedient than lengthier blog posts, press releases or articles.

(3) Coca-Cola Germany was named 'Most Proactive Web 2.0 Company' last week at IIR Technology Web 2.0 Congress! Good job, Germany! ^OP 5:06 PM May 3rd via CoTweet.

(4) Coca-Cola Singapore recently introduced 3 new drinks-Ice Lemon Tea, Ice Apple Tea, and Ice Passionfruit Tea! YUMMM! ^OP 4:52 PM May 4th via CoTweet.

(5) We are joining forces with Gov. Schwarzenegger to announce 8 new 'Live Positively' fitness centers in California! ^OP 3:15 PM May 6th via CoTweet.

(6) Taiwanese pop singer A-Mei is the official endorser for Open Happiness in Taiwan. Her song is awesome! ^OP 9:46 AM May 25th via CoTweet.

In terms of structure, the first element is the message proper, which is then followed by the timestamp generated by the Twitter template and the indication of the medium underlined, hyperlinked and highlighted in a different colour (CoTweet and TwitPic here, but several other options exist, like Bitty, Facebook, HootSuite and TweetButton). In the specific case of @CocaCola (unlike @WholeFoods, for example), the task of embodying the collective voice of the company is distributed among different team members, so the initials of the author of a given tweet appear before the timestamp (for example, GD, JW, LF and OP). Besides the informal register typical of advertising and characterised by positive evaluation and euphoric exclamations, the other perceivable features typical of conversational digital genres are: "relaxed syntax and punctuation, use of paralinguistic expressions, […] onomatopoeia, emoticons, and playful tropes" (Erickson 2000: 3). In line with the feeling of immediacy and the pre-eminence of the present moment which are typical of Twitter, verb tenses are usually the present simple or continuous, more rarely the future, and the imperative for invitations, recommendations and wishes. When the past tense or the present perfect occur, the task of making close temporal proximity explicit falls on time adverbials (e.g. *just, last week, recently*).

In the sample analyzed here, we observe that updates are much fewer than replies, a strategy which is applicable by inference to the entire CocaCola timeline well beyond the one-month time period taken into consideration for this research and which is consistent with the decision to give room to consumers. The basic moves at work are always promoting company happenings and engaging an online discourse community of loyal brand fans, who represent "individuals with well-established attitudes" and "specific brand beliefs" towards the product (Drolet/Aaker 2002: 61) according to the analytic framework of consumer psychology.

The recurrent topics of the updates construct the company as a truly global, environmentally friendly and socially responsible enterprise, as shown in the examples below.

- Company development

(7) Coca-Cola System to invest $300 Million in Pakistan! Hellllooo Pakistan. ^OP 12:53 PM May 22nd via CoTweet.

(8) Coca-Cola Brasil Maceió plant is the first soft drink green production unit in Latin America. ^OP 7:25 AM May 12th via CoTweet.

- Sustainability and green living

(9) Coca-Cola China launches new lightweight bottle 4 Ice Dew water. It contains 35% less PET & can be easily compressed prior to recycling. ^OP 10:01 AM May 8th via CoTweet.

(10) 400 associates at juice plant in FL stepped up & enlisted volunteers 2 help rescue birds following oil spill. http://CokeURL.com/2255 ^OP 6:57 AM May 26th via CoTweet.

- Sponsorship

(11) Coke Brasil is sponsoring the 2010 Parintins Festival! Here is some more info on the history of the festival: http://CokeURL.com/vzrs ^OP 6:58 AM May 5th via CoTweet.

(12) We are partnering w/ Goal.com to count down the most memorable World Cup celebrations of all time! Check it out! ^OP 1:06 PM May 10th via CoTweet.

- Health

(13) Healthy Weight Commitment Foundation unveils 3 new health sites: energybalance101.com, parents.com/hwcf, and yourwellnessadvantage.com! ^OP 7:24 AM May 8th via CoTweet.

(14) The Coca-Cola Company & 15 other food & beverage manufacturers have pledged 2 cut 1.5 trillion calories from products by the end of 2015 ^OP 11:58 AM May 20th via CoTweet.

Replies differ from updates, as their aim is to provide a 'one-to-one' response in what is normally a minimal turn-taking situation, in which dyadic exchanges often stop after a first reply. They begin with the explicit notation of the recipient's username as registered on the social network. The username is preceded by the @ sign as a marker of addressivity (Werry 1996) and followed by the text of the message, the timestamp, the channel, the set phrase *in reply to* and the repetition of the recipient's username. Listed in the profile timeline in reverse chronological order together with status updates, replies are shared by all the followers of that account, i.e. all ratified participants in Goffman's words (1981). As they are by far the majority of tweets, replies make up the bulk of the Coca-Cola profile timeline, their main communicative purpose being that of keeping the conversation going between the company (indexed by the use of the inclusive *we*) and its followers. With their bland and fragmented wording, they can be described as instances of "phatic communion where utterances with unexceptional content are used to signal communicative connection" (Page 2010: 426). The fact that the Coca-Cola conversation does not pursue debates on popular topics would seem to be further demonstrated by the absence of 'hashtags' in the timeline, i.e. key words or phrases prefixed by a hash symbol (#) that help to keep track of common news items or recurrent themes, making it easier to find conversational threads in a search.

(15) @crushonshan Hi Shan, we love all of our Coca-Cola fans! Have a great day! ^LF 5:31 AM May 1st via CoTweet in reply to crushonshan.

(16) @Meg9190 Meghan, thanks for preferring always Coca-Cola! ^GD 9:53 AM May 4th via CoTweet in risposta a Meg9190.

(17) @CarlaMichelle95 We can understand that addiction, Carla. It's hard to resist an ice cold, refreshing Coca-Cola. ^JW 12:45 PM May 25th via CoTweet in reply to CarlaMichelle95.

In terms of move structure, replies alternate between greeting, thanking ("we thank you for your loyalty", "thanks for the tweet/retweet", "thanks for sharing", "we're happy to follow you"), congratulating, appreciating customers' loyalty ("we appreciate the retweet", "Yes we appreciate our loyal fans like you"), offering assistance, at times apologising and, ultimately, promoting the brand, as in the three examples below.

(18) @alyssaavant Alyssa, thanks Alyssa. We look forward to assisting you. ^GD 10:04 AM May 11th via CoTweet in reply to alyssaavant.

(19) @BellandaInParis Well thanks for being a loyal fan, Bellanda! Enjoy your Coca-Cola! ^OP 5:42 PM May 18th via CoTweet in reply to BellandaInParis.

(20) @whatsername_x3 Thanks for the tweet, Tinker Bell! You're right, Coca-Cola is truly awesome! ^LF 10:10 AM May 25th via CoTweet in reply to whatsername_x3.

By clicking on the hyperlink of the username, it is possible to retrieve the tweets that generated the above replies (examples 21, 22 and 23).

(21) @CocaCola I filled in the form. Thank you! 10:01 AM May 11th via TweetDeck in reply to CocaCola.

(22) @CocaCola Wow! Thanks for following back (she says while holding a Coca Cola in hand) ;) Yum! 1:09 PM May 18th via web.

(23) I love drinking @CocaCola. It's awesome. :D 9:46 AM May 25th via web.

As examples from 13 to 18 show, a recurrent linguistic strategy of replies is that of embedding the user's selected profile name within the message ("Shan", "Meghan", "Carla", "Alyssa", "Bellanda" and "Tinker Bell"). Arguably, the function of naming is to work as a cohesive device at the linguistic level, while rhetorically re-enforcing a personalized relationship with consumers by involving them in the co-construction of the brand.

Despite its obvious redundancy the other lexical item endowed with a comparable cohesive function is the brand name in full – "Coca-Cola" – in examples 13, 15, 17 and 18. The rhetorical strategy of naming the brand on any possible occasion is clearly borrowed from advertising and should be understood in this context as a form of manipulation through corporate control of discourse, now occurring even more pervasively than in the past at the level of global communications (Fairclough 2003: 99-111). Intertextual references to successful payoffs of previous ad campaigns are also detectable in the Twitter stream. A "refreshing" "ice cold Coca-Cola", the "Coke side of life", the "Real Thing", "Enjoy" used intransitively – "Enjoy your Coke" – and "Open Happiness" are all citations gleaned from past advertising slogans,[4] memorized in the archives of popular culture and re-used for branding on Twitter.

(24) @tazuco Hector, YES! an ice cold Coca-Cola always makes our days much happier and much nicer! Enjoy. ^GD 9:39 AM May 10th via CoTweet in reply to tazuco.

(25) @aolpersonals aolpersonals, YAY! any excuse to drink a Coca-Cola is always a refreshing excuse! Enjoy. ^GD 9:06 AM May 10th via CoTweet in reply to aolpersonals.

---

4 More specifically the slogans are: "Whoever you are, whatever you do, wherever you may be, when you think of refreshment, think of ice cold Coca-Cola" (1939); "It's the real thing" (1969); "Always Coca-Cola" (1993); "Enjoy" (2000); "The Coke side of life" (2006); "Open Happiness" (2009).

(26) @nunz93 Thanks for your loyalty! We're glad you are enjoying a refreshing Coca-Cola beverage. JW 6:34 PM May 13th via CoTweet in reply to nunz93.

(27) @Karliux86 Karla, glad to have you on the Coke Side of Life. ^GD 6:33 AM May 4th via CoTweet in reply to Karliux86.

(28) @AMERICANFREEDOM JAV, yes, nothing like The Real Thing! an ice cold Coca-Cola. ^GD 3:38 AM May 6th via CoTweet in risposta a AMERICANFREEDOM.

(29) @ebasedBOB wonderful, Bob. Always Coca-Cola. ^GD 10:08 AM May 24th via CoTweet in reply to ebasedBOB.

(30) @HelloShiibi Shiibi, have a wonderful day and enjoy your Coke. Thx for the tweet! ^LF 10:44 AM May 8th via CoTweet in reply_to HelloShiibi.

(31) @MileysLipgloss Hi Vanessa, I am in Atlanta, doing great and opening lots of happiness! ^GD 7:53 AM May 18th via CoTweet in reply_to MileysLipgloss.

In other words, despite their apparently offhand character, replies creatively manage to rework user-generated content from loyal customers' tweets together with past taglines. Thus, out of user feedback and the standardized formulas of advertising, they mould what can be regarded as *customized* instances of promotional discourse, targeted at individual consumers, which expand the entire semiotic universe generated by the brand. The often utmost banality of the ongoing conversation manages nonetheless to establish, keep up and expand networks.

Such ubiquitous branding, realized by hybridizing diverse generic resources, lies at the basis of the strategy to use micro-blogging to further and enhance the conversation carried on in other sectors of the Company's website, whose aim is to promote a holistic perception of the product as a choice of lifestyle obtained through faithful consumption. A very good example of this synergy is offered by the Coca-Cola Conversations blog also embedded on the company website.

(32) *About this Coca-Cola Blog*
Welcome to the Coca-Cola Conversations blog! We'll share information on a wide variety of topics, ranging from our role in history to Coke collectibles. However, the blog only works if there is a two-way dialogue. We look forward to chatting with you!

The blog posts are primarily about items from the Coca-Cola collection, or stories about the company and its brands, in which the dominant topos is that of heritage. "Key events in Company history" and the "contextual framework around them" are provided, with the information presented in an informative and entertaining manner" and "establishing links that demonstrate the connections between heritage and contemporary business practice" (Mooney 2009). In other words, blog posts well typify the hybrid nature of multi-purpose advertising as a facet of branding, in which information, education and promotion overlap.

(33) 1936 China Paper Poster at the World of Coke
You have probably heard that the Expo 2010 in Shanghai is opening tomorrow. In honor of the opening, I'd like to spotlight an interesting China related poster on display at the World of Coke. This 1936 poster features a fashionable woman holding a glass of Coke, right under a Chinese Coca-Cola logo. The interesting thing about this piece is that while it clearly appears to be a piece of foreign advertising, it was actually produced for use in American Chinatowns. For this reason, it is often referred to in collectors' circles as "Chinatown". This can be found in the Milestones of Refreshment case in the section highlighting fashion. (April 30)

Quite interestingly, the mention of this vintage item – the 1936 China Paper Poster – is taken up again in an update on Twitter a few days later after the online publication of the post (example 34), which shares the related link with the entire community.

(34) 1936 China Paper Poster at the @worldofcoke via @coke_archives: http://CokeURL.com/6cxw ^OP 9:36 AM May 5th via CoTweet.

Other heritage-related events are also advertised on the social network (examples 35 and 36), after being previously or repeatedly dealt with on the corporate blog (example 37 for Jacobs' Pharmacy in Atlanta, the birthplace of Coca-Cola, and examples 38 and 39 for Orange Fanta).

(35) Today is National Have A Coke Day! Also – today is 124 anniversary of the day the 1st glass of Coke was served @ Jacobs' Pharmacy in ATL. ^OP 8:54 AM May 8th via CoTweet.

(36) Did you know: Fanta Orange was introduced in Italy in 1956. ^OP 4:21 PM Apr 30th via CoTweet.

(37) *Did you know? 1886 vs. today*
On May 8, 1886, a pharmacist named Dr. John Pemberton introduced Coca-Cola at Jacobs' Pharmacy in downtown Atlanta, where it sold for 5 cents a glass. (January 23, 2008)

(38) *Happy Birthday, Fanta!*
We mark June 1st as the date that Fanta was introduced to the US market and even blogged about it a few years ago. (June 1, 2010)

(39) *Fanta Orange hits the U.S.*
Forty-eight years ago this month, Fanta hit store shelves in the United States. (June 18, 2008)

Like the intertextual references to past ad campaigns, these thematic correspondences show how Twitter reworks the hybrid content of other promotional genres, adapting it to its generic constraints and its networks.

4. Concluding remarks

Taking the world's most famous soft drink as a case study, the analysis has approached online advertising via Twitter in the recognition that the social network is increasingly being integrated into corporate websites for branding purposes. The analytic approach here pursued has underlined the connection between the codification of genres and societal changes, in particular new media communication and transformations in consumption patterns. These changes witness the emergence of a new breed of consumers, actively involved in brand promotion, whose communities aggregate on social networks like Twitter. As an emerging advertising genre with a marked participatory nature, corporate tweets would seem to express these recent social configurations that could be taken as an expression of the *agora*, that is, of the public at large and not just of a selected professional elite, according to the interesting metaphor advanced by Berkenkotter and Huckin (1995).

First, attention has been paid to the linguistic and discursive encoding of tweets with regard to their effectiveness as persuasive appeals within quite strict generic constraints, especially their conciseness. The analysis has shown that, far from being naïve, their communicative economy, realized by means of an oversimplification of linguistic features and formulaic rhetorical strategies, manages to customize promotional messages in line with the underlying corporate philosophy. This unrelenting branding attitude affects the entire move structure of tweets, both when announcing company happenings in status updates and when replying to customers' messages.

A second remark that can be derived from the above considerations concerns genre hybridization within the online environment. The sample analyzed here demonstrates that new text types like tweets do not supersede but rather co-exist side by side with previous advertising genres, borrowing from them and enhancing their mutual role, while maintaining their connoting features. The phenomenon of genre hybridization used to empower persuasive appeals is also well apparent in the net of intertextual references that link tweets to past advertising slogans and micro-blogging to blogging on the same website.

Preliminary observations suggested that the creation of a strategic synergy between the producing and receiving ends of the market has resulted in the "inclusion of customer's everyday life within the branded context supplied by the provider" (Arvidsson 2006: 116), facilitated by the development of 'social networks' "as a kind of secondary layer to ordinary life" (Arvidsson 2006: 177). Indeed, incorporating the voices of the *agora* appears as the most recent trend of ubiquitous branding via Twitter, which sees the technology of mobile connectedness turn into an unprecedented form of customized branding. However, when weighed against the surprising affordances of social networking, the banal and largely acritical content of most tweets still raises suspicions about the overall quality of the social capital thus circulated and about the meaningfulness of consumer participation, swinging between proactive engagement and ubiquitous surveillance.

References

Arvidsson, Adam 2006. *Online Branding: Meaning and Value in Media Culture*. London and New York: Routledge.

Banks, John / Potts, James 2010. Co-creating Games: A Co-evolutionary Analysis. *New Media & Society* 12/2, 253-270.

Bendandi, Michelangelo 2009. Should CEOs Twitter? *Brunswick Review*. 2, 22-23. Available at <http://www.brunswickgroup.com>.

Berkenkotter, Carol / Huckin, Thomas N. 1995. *Genre Knowledge in Disciplinary Communication: Cognition/Culture/Power*. Hillsdale, NJ: Lawrence Erlbaum Associates.

Bhatia, Vijay K. 1993. *Analysing Genre: Language Use in Professional Settings*. London: Longman.

Bhatia, Vijay K. 2000. Genres in Conflict. In Trosborg, Anna (ed.) *Analysing Professional Genres*. Amsterdam/Philadelphia: John Benjamins, 147-161.

Bhatia, Vijay K. 2002. Applied Genre Analysis: A Multi-Perspective Model. *Ibérica* 4, 3-19.

Bhatia, Vijay K. 2004. *Worlds of Written Discourse: A Genre-Based View*. London: Continuum.

Boyd, Danah M. / Ellison, Nicole B. 2007. Social Network Sites: Definition, History, and Scholarship. *Journal of Computer-Mediated Communication* 13/11. Available at <http://jcmc.indiana.edu/vol13/issue1/boyd.ellison.html>.

Drolet, Aimee/Aaker, Jennifer 2002. Off-Target? Changing Cognitive-Based Attitudes. *Journal of Consumer Psychology* 12/1, 59-68.

Erickson, Thomas 2000. Making Sense of Computer-Mediated Communication (CMC): Conversations as Genres, CMC Systems as Genre Ecologies. *Proceedings of the 33rd Hawaii International Conference on System Sciences*, 1-10. Available at <http://doi.ieeecomputersociety.org/10.1109/HICSS.2000.926694>.

Fabris, Giampaolo 2008. *Societing. Il marketing nella società postmoderna*. Milan: Egea.

Fairclough, Norman 2003. *Analysing Discourse: Textual Analysis for Social Research*. London: Routledge.

Goffman, Erving 1981. Footing. In *Forms of Talk*. Philadelphia: University of Pennsylvania Press, 124-159.

Herring, Susan C. / Scheidt, Lois Ann / Bonus, Sabrina / Wright, Elijah 2004. Bridging the Gap: A Genre Analysis of Weblogs. *Proceedings of the 37th Hawaii International Conference on System Sciences*, 1-11. Available at <http://doi.ieeecomputersociety.org/10.1109/HICSS.2004.1265271>.

Honeycutt, Courtenay / Herring, Susan C. 2009. Beyond Microblogging: Conversation and Collaboration via Twitter. *Proceedings of the 42nd Hawaii International Conference on System Sciences*, Los Alamitos, CA: IEEE Computer Society Press, 1-10. Available at <http://www.computer.org/portal/web/csdl/doi/10.1109/HICSS.2009.602>.

Israel, Shel 2009. *Twitterville: How Businesses Can Thrive in the New Global Neighbourhoods*. New York: Portfolio.

Jackson, Anna / Yates, JoAnne / Orlikowski, Wanda 2007. Corporate Blogging: Building Community through Persistent Digital Talk. *Proceedings of the 40th Hawaii International Conference on System Sciences.* Los Alamitos, CA: IEEE Computer Society Press, 1-10. Available at <http://citeseerx.ist.psu.edu/viewdoc/summary?doi=10.1.1.104.7371>.

Jenkins, Henry 2006. *Convergence Culture: Where Old and New Media Collide*. New York and London: New York University Press.

Koteyko, Nelya 2009. 'I Am a Very Happy, Lucky Lady, and I Am Full of Vitality!' Analysis of Promotional Strategies on the Websites of Probiotic Yoghurt Producers. *Critical Discourse Studies* 6/2, 111-125.

Kuwabara, Kazuhiro / Watanabe, Takumi / Ohguro, Takeshi / Itoh, Yoshihiro / Maeda, Yuji 2002. Connectedness Oriented Communication: Fostering a Sense of Connectedness to Augment Social Relationships. *IPSJ Journal* 43/11, 3270-3279.

Mooney, Philip 2009. The Blog as an Archival Tool: Coca-Cola Conversations. *The Interactive Archivist.* 22 June. Available at <http://lib.byu.edu/sites/interactivearchivist>.

Page, Ruth 2010. Re-examining Narrativity: Small Stories in Status Updates. *Text & Talk* 30/4, 423-444.

Schroeder, Ralph 2010. Mobile Phones and the Inexorable Advance of Multimodal Connectedness. *New Media & Society* 12/1, 75-90.
Schumann, David W. / Thorson, Esther 2007. *Internet Advertising: Theory and Research.* Mahwah, NJ: Lawrence Erlbaum Associates.
Simpson, James 2005. Conversational Floors in Synchronous Text-based CMC Discourse. *Discourse Studies* 7/3, 337-361.
Starling, David Hunter III. 2009. Structural Holes and Click-Throughs in Online Advertising. *Social Science Research Network*, 1-32- Available at <papers.ssrn.com/sol3/papers.cfm?abstract_id=1457569>.
Toffler, Alvin 1980. *The Third Wave*. London: Pan Books.
Watkins, Thomas G. 2009. From Twitter to Advertising Treasure. *CNN.com*, Nov. 24. Available at <http://www.cnn.com/2009/TECH/11/24/twitter.ads/index.html>.
Werry, Cristopher C. 1996. Linguistic and Interactional Features of Internet Relay Chat. In Herring, Susan C. (ed.) *Computer-mediated Communication: Linguistic, Social and Cross-Cultural Perspectives.* Amsterdam: John Benjamins, 47-63.

Websites

The Coca-Cola Company <http://www.thecoca-colacompany.com>.
The Coca-Cola Conversations blog <http://www.coca-colaconversations.com>.
CoTweet <http://cotweet.com>.

Notes on Contributors

SANDRA CAMPAGNA is Associate Professor of English Language and Translation at the University of Turin, Faculty of Economics. Her principal research areas are translation studies, crosscultural studies, sociolinguistics and intercultural communication in economic discourse and the web language domain (*Discoursal Strategies On-line: an Intercultural Approach to the Language of Charities*, 2004). She is currently working on identity issues and multimodal/multimedial discourse. She has co-edited with Martin Solly and Michelangelo Conoscenti the volume *Verbal/Visual Narrative Texts in Higher Education* (2008) and has co-authored with Cecilia Boggio the volume *Multimodal Business and Economics* (2009). Her recent publications include 'Projecting Visual Reasoning in Research Conference Presentations', in Maurizio Gotti (ed.) *Commonality and Individuality in Academic Discourse* (2009) and 'Semiotic Change and Identity in Corporate Culture Representations', in Gabriella Di Martino / Vanda Polese / Martin Solly (eds) *Identity and Culture in English Domain-specific Discourse* (2009).

PAOLA CATENACCIO, PhD (Catholic University Milan), MA (Birkbeck College, London) is Associate Professor of English Linguistics and Translation at Università degli Studi di Milano, where she teaches English Linguistics courses for the degree programme in Linguistic and Cultural Mediation. Her research interests span from second language acquisition to translation and discourse analysis, especially in the context of LSP research. Her most recent contributions focus on the genre of the press release and, more in general, on linguistic aspects of corporate communication. Some of her latest publications are: *Corporate Press Releases: An Overview* (2008); "De-humanising the alien: The construction of migrants' rights in EU legislation" (2007); "Constructing identities in the fashion industry: Building brand and customer image through press releases" (2007); "Press releases as a hybrid genre: Addressing the informative/promotional conundrum" (2008). She has also co-edited, with Giuliana Garzone and Gina Poncini, *Multimodality in Corporate Communication. Web Genres and Discursive Identity* (2007) and, with Giuliana Garzone, *Language and Bias in Specialised Discourse* (2008).

ELISA CORINO finished her PhD in "Linguistics, Applied Linguistics and Linguistic Engineering" in 2008, she is now Research fellow at the Department of Literary and Philological Science – University of Turin. She teaches academic courses on Teaching of Modern Languages. Her research areas include Italian as L2/LS, Computer Mediated Communication, corpus linguistics and translation studies. At present her work concentrates especially on corpus analysis of formal varieties in web language.

CHIARA DEGANO is a Tenured Researcher in English Linguistics and Translation at Università degli Studi di Milano, Italy. Her research is centred on Discourse Analysis, which she has integrated with the quantitative approach of corpus linguistics and with aspects of Argumentation theory. Within this framework she has pursued a twofold line of research, focusing on the representation of international conflicts in the media (*Discorsi di guerra: il prologo del conflitto iracheno sulla stampa britannica e italiana,* 2008; "Rhetoric and bias in Editorials: the Lebanese crisis of 2006 in the British press", 2008, with G. Garzone; "Presupposition and dissociation in discourse: a corpus study", 2007), and on professional discourse, which she investigated in the perspective of ideological and evaluative elements, with specific reference to issues of identity ("The construction of cultural identity in banks' reports on corporate social responsibility", in press). She has also taken part in an international project on arbitration, investigating generic tensions in international awards. Her other research interests fall in the area of translation and interference, with a concern for syntactic interference in translations from English into Italian and for the rendering of multi-cultural references in film dubbing.

GIULIANA GARZONE is Full Professor of English Linguistics and Translation at Università degli Studi di Milano, Italy. Her main field of investigation is specialized discourse in its different domains, which she has explored in a discourse analytical perspective, integrating it whenever necessary with computerized interrogation routines, also in light of her work on the theoretical issues involved in combining Critical Discourse Analysis with Corpus Linguistics ("What can corpus linguistics do for Critical Discourse Analysis?", 2004, with F. Santulli). She has co-ordinated several research projects and published extensively on legal language (*Performatività e linguaggio giuridico: una proposta di classificazione*, 1999; *Il linguaggio giuridico. Prospettive interdisciplinari*, 2008, with F. Santulli), on scientific discourse (*Perspectives on ESP and Popularization*, 2006), and on business communication. In the latter domain, she has recently devoted specific attention to multimodality and to the analysis of text types and genres embedded in the web-mediated environment ("Genres, multimodality and the World-Wide Web: Theoretical Issues", 2007), an issue she has also investigated with reference to other types of communication, e.g. political communication (cf. "Public Diplomacy, Multimodality and The World-Wide Web", with C. Degano). She has also (co-)edited several volumes (most recently, *Discourse, Ideology and Specialized Communication*, with S. Sarangi, 2007; *The Use of English in Institutional and Business Settings,* with C. Ilie, 2007; *Identities across Media and Modes: Discursive Perspectives,* with P. Catenaccio, 2009; *Discourse and Identities in Institutional, Professional and Academic Communication,* with J. Archibald, 2010). Her other research interests are in translation and interpreting studies, on which she has also published extensively (among others: *Interpreting in the 21st Century,* with M. Viezzi, 2002; *Esperienze del tradurre*, 2005). She is general editor (with A. Cardinaletti e F. Frasnedi) of the book series *Lingua, traduzione, didattica.*

MARISTELLA GATTO is researcher of English Language and Translation at the University of Bari (Italy). Her main areas of research include Corpus Linguistics (especially the Web as Corpus), Translation Studies, and Tourism Discourse. She is the author of a monograph and articles on the renegotiation of the notion of corpus under the impact of the web (*From Body to Web. An Introduction to the Web as Corpus*, Laterza 2009; *The body and the web. The web as corpus 10 years on*, Icame Journal 2011). She has coedited the volume *Translation.*

The State of the Art / La Traduzione. Lo Stato dell'Arte (2007) and published essays on literary translation (including Seamus Heaney's translation of *The Divine Comedy*, Richard Fanshawe's translation of Guarini's *Pastor Fido* and early translations of Manzoni's *I Promessi Sposi*). She is also author of the volume *Dublin. From paralysis to international tourism* (2007) and of articles on translation and representation in tourism discourse.

ENRICO GRAZZI is a researcher in English Language and Translation at the University of 'Roma Tre', Italy, Department of Linguistics, where he teaches English for Communication Sciences. Main interests: English as a lingua franca; educational linguistics; sociocultural theory; corpus linguistics. His main publications associated with these areas of research are: "Language transfer revisited. The Global English perspective and its pedagogic implications", in *EIL, ELF, Global English: Teaching and Learning Issues*, Cesare Gagliardi & Alan Maley (eds, 2010); "Which English? Whose English?", in *L'Italiano errante – n. 6, Lingue e Culture fra Identità e Potere*, 2009; "A corpus-based analysis of modality in four English textbooks for the First Certificate Exam", in the conference proceedings of AIA's XXIII National Convention, Vol. 3 (2009); "Attivazione di un sistema tutoriale. Predisposizione e valutazione del materiale didattico di supporto al tutorato per la lingua inglese, livello A2", in *Dalla scuola all'università: una scelta di vita* (2008); *Horizons on Literature. An Approach to 19th and 20th Century Literature* (2006). He was President of TESOL-Italy (2002-2004). Recipient of a scholarship from the British Council and MIUR, in 1999 he qualified as teacher trainer at the University of East Anglia and Norwich Institute for Language Education (NILE), Norwich (UK).

CORNELIA ILIE is Professor of English Linguistics at Malmö University, Sweden. Her research interests, which are largely interdisciplinary, focus on Pragmatics, Institutional Discourse Analysis, Intercultural Rhetoric and Argumentation Theory. Cornelia Ilie is the founder and president of ESTIDIA (European Society for Transcultural and Interdisciplinary Dialogue) and an associate member of the Centre of Excellence in Political Thought and Conceptual Change (COE PolCon) of the Jyväskylä University, Finland. She is Board member of *IADA*

(International Association for Dialogue Analysis), member of the IPrA (International Pragmatics Association) Consultation Board, and member of the *IAU* (International Association of Universities) Reference Group of Experts on Higher Education and Education For All. She has published extensively (encyclopedia entries, refereed journal articles and book chapters) on institutional discourses (particularly political, media and academic discourses), cross-cultural communication strategies, rhetoric and argumentation in dialogue. Among her representative publications worth mentioning are *Language and Ideology,* Vol. 2: *Descriptive Cognitive Approaches* (2001), *Language, Culture, Rhetoric* (2004), *The use of English in Institutional and Business Settings: An Intercultural Perspective* (2007) and *European Parliaments under Scrutiny* (2010). She is member of the editorial board of a range of international journals and of the book series *Argumentation in Context.*

BETTINA MOTTURA, PhD, is a Tenured Researcher in the Department of Contemporary Languages and Cultures of Università Statale di Milano (Italy). Her research concentrates on Contemporary Chinese language and culture, with a focus on the language in use in the institutions of the People's Republic of China. In particular, in an earlier stage of her research, she investigated the reform of communication and linguistic strategies in public administration. More recently, she has turned her attention to the definition of genres in institutional communication and their evolution with the aim of tracking social and political changes both within the professional community and in the relationship between institutional actors and the general public. At present her work concentrates on online materials posted on official government websites. Her recent publications include "The Changing Terminology of Chinese Public Administration" (2006); Administrative Specialized Language Training in China and in Italy Today: in Search of Renewed Professional Identities for Civil Servants (2007); Chinese civil servants and the creation of a discourse community to promote social change (2010),

CRISTINA ONESTI holds a PhD in Linguistics and Applied Linguistics and she is now Research fellow at the University of Turin, Department of Literary and Philological Science. Her research areas include corpus linguistics, Computer Mediated Communication and language teaching. In particular, her work concentrates now on corpus analysis of web language. She teaches academic courses on Applied Linguistics and Italian as L2. She has participated in recent European Projects focused on Teacher Training.

MARIA CRISTINA PAGANONI, PhD, is a tenured researcher in English Language and Translation at Università degli Studi di Milano. Her current research focuses on web-mediated communication and digital genres, with the avail of an interdisciplinary approach that draws on discourse analysis, multimodality and social semiotics. Her recent publications include the following articles and book chapters: "Tsunami and Money: Humanitarian Aid in Media Coverage of the Asian Catastrophe" (2008), "'The Opinion and the Counter Opinion': News Framing and Double Voicing on Al Jazeera English" (2009), "From Hyde Park to the Planetary Garden: Rhetorics of Development at the London 1851 and Milan 2015 World Exhibitions" (2010) and "Food for Thought: The Discursive Rehearsal of the Milano Expo 2015" (2011). At present she is working on a publication on web-based city branding, which stems from her fascination with representational strategies, public diplomacy and active citizenship, as well as the role of major promotional events such as international exhibitions in the cultural economy of the global city in the Information Age.

ELIZABETH ROWLEY-JOLIVET is a former Senior Lecturer in English for Specific Purposes at the École Polytechnique, University of Orléans in France. She is a member of the Laboratoire Ligérien de Linguistique and of the professional association Mediterranean Editors and Translators. Her research interests cover genre analysis, spoken and written communication in science, multimodality and the epistemology of science. She has built up a large oral corpus of filmed scientific conference presentations, which has formed the basis for a number of international publications on visual discourse in science, disciplinary textual patterns, the construction of scientific facts, scientific rhetoric and syntactic analysis. Her current research focuses on the specific problems of French-speaking researchers publishing or communicating in English, and on the evolution of genres in the Web 2.0 era.

MAŁGORZATA SOKÓŁ, PHD, is Assistant Professor in the Department of English at Szczecin University, Poland. Her major fields of professional interest include: computer-mediated communication and computer-mediated discourse analysis, professional and institutional discourse analysis, genre theory, corpus linguistics, and recently, digital humanities. She has researched into aspects of digital genre development and evolution in professional and organisational communication, and investigated patterns of identity construction and evaluative language use in digital genres. She edited *Perspectives on Othering and Stereotyping: Language, Literature, Culture* (2010), co-edited *Us and Them – Them and Us: Constructions of the Other in Cultural Stereotypes* (2011) and authored *Discoursal Construction of Academic Identity in Cyberspace: The Example of An E-Seminar* (2012).

ALESSANDRA VICENTINI received a degree in History of the English Language and a doctoral degree in English Language Studies from the University of Milan (Italy). She is currently a tenured researcher in English Linguistics and Translation at the Faculty of Mathematics, Physics and Natural Sciences of the University of Insubria, Varese (Italy), where she teaches English Linguistics and English for Scientific Purposes. Her research interests include English and Anglo-Italian grammaticography and lexicography (XVI-XVIII centuries), Italian/English contrastive linguistics, English for Specific Purposes (medical-scientific language, in particular), Discourse Analysis, Web Communication Theory and Multimodality.

Linguistic Insights

Studies in Language and Communication

This series aims to promote specialist language studies in the fields of linguistic theory and applied linguistics, by publishing volumes that focus on specific aspects of language use in one or several languages and provide valuable insights into language and communication research. A cross-disciplinary approach is favoured and most European languages are accepted.

The series includes two types of books:

- **Monographs** – featuring in-depth studies on special aspects of language theory, language analysis or language teaching.
- **Collected papers** – assembling papers from workshops, conferences or symposia.

Each volume of the series is subjected to a double peer-reviewing process.

Vol. 1 Maurizio Gotti & Marina Dossena (eds)
Modality in Specialized Texts. Selected Papers of the 1st CERLIS Conference.
421 pages. 2001. ISBN 3-906767-10-8 · US-ISBN 0-8204-5340-4

Vol. 2 Giuseppina Cortese & Philip Riley (eds)
Domain-specific English. Textual Practices across Communities and Classrooms.
420 pages. 2002. ISBN 3-906768-98-8 · US-ISBN 0-8204-5884-8

Vol. 3 Maurizio Gotti, Dorothee Heller & Marina Dossena (eds)
Conflict and Negotiation in Specialized Texts. Selected Papers of the 2nd CERLIS Conference.
470 pages. 2002. ISBN 3-906769-12-7 · US-ISBN 0-8204-5887-2

Vol. 4 Maurizio Gotti, Marina Dossena, Richard Dury, Roberta Facchinetti & Maria Lima
Variation in Central Modals. A Repertoire of Forms and Types of Usage in Middle English and Early Modern English.
364 pages. 2002. ISBN 3-906769-84-4 · US-ISBN 0-8204-5898-8

Editorial address:

Prof. Maurizio Gotti Università di Bergamo, Facoltà di Lingue e Letterature Straniere,
Via Salvecchio 19, 24129 Bergamo, Italy
Fax: 0039 035 2052789, E-Mail: m.gotti@unibg.it

Vol. 5 Stefania Nuccorini (ed.)
Phrases and Phraseology. Data and Descriptions.
187 pages. 2002. ISBN 3-906770-08-7 · US-ISBN 0-8204-5933-X

Vol. 6 Vijay Bhatia, Christopher N. Candlin & Maurizio Gotti (eds)
Legal Discourse in Multilingual and Multicultural Contexts.
Arbitration Texts in Europe.
385 pages. 2003. ISBN 3-906770-85-0 · US-ISBN 0-8204-6254-3

Vol. 7 Marina Dossena & Charles Jones (eds)
Insights into Late Modern English. 2nd edition.
378 pages. 2003, 2007.
ISBN 978-3-03911-257-9 · US-ISBN 978-0-8204-8927-8

Vol. 8 Maurizio Gotti
Specialized Discourse. Linguistic Features and Changing Conventions.
351 pages. 2003, 2005.
ISBN 3-03910-606-6 · US-ISBN 0-8204-7000-7

Vol. 9 Alan Partington, John Morley & Louann Haarman (eds)
Corpora and Discourse.
420 pages. 2004. ISBN 3-03910-026-2 · US-ISBN 0-8204-6262-4

Vol. 10 Martina Möllering
The Acquisition of German Modal Particles. A Corpus-Based Approach.
290 pages. 2004. ISBN 3-03910-043-2 · US-ISBN 0-8204-6273-X

Vol. 11 David Hart (ed.)
English Modality in Context. Diachronic Perspectives.
261 pages. 2003. ISBN 3-03910-046-7 · US-ISBN 0-8204-6852-5

Vol. 12 Wendy Swanson
Modes of Co-reference as an Indicator of Genre.
430 pages. 2003. ISBN 3-03910-052-1 · US-ISBN 0-8204-6855-X

Vol. 13 Gina Poncini
Discursive Strategies in Multicultural Business Meetings.
2nd edition. 338 pages. 2004, 2007.
ISBN 978-3-03911-296-8 · US-ISBN 978-0-8204-8937-7

Vol. 14 Christopher N. Candlin & Maurizio Gotti (eds)
Intercultural Aspects of Specialized Communication.
2nd edition. 369 pages. 2004, 2007.
ISBN 978-3-03911-258-6 · US-ISBN 978-0-8204-8926-1

Vol. 15 Gabriella Del Lungo Camiciotti & Elena Tognini Bonelli (eds)
Academic Discourse. New Insights into Evaluation.
234 pages. 2004. ISBN 3-03910-353-9 · US-ISBN 0-8204-7016-3

Vol. 16 Marina Dossena & Roger Lass (eds)
Methods and Data in English Historical Dialectology.
405 pages. 2004. ISBN 3-03910-362-8 · US-ISBN 0-8204-7018-X

Vol. 17 Judy Noguchi
The Science Review Article. An Opportune Genre in
the Construction of Science.
274 pages. 2006. ISBN 3-03910-426-8 · US-ISBN 0-8204-7034-1

Vol. 18 Giuseppina Cortese & Anna Duszak (eds)
Identity, Community, Discourse. English in Intercultural Settings.
495 pages. 2005. ISBN 3-03910-632-5 · US-ISBN 0-8204-7163-1

Vol. 19 Anna Trosborg & Poul Erik Flyvholm Jørgensen (eds)
Business Discourse. Texts and Contexts.
250 pages. 2005. ISBN 3-03910-606-6 · US-ISBN 0-8204-7000-7

Vol. 20 Christopher Williams
Tradition and Change in Legal English. Verbal Constructions in Prescriptive Texts.
2nd revised edition. 216 pages. 2005, 2007. ISBN 978-3-03911-444-3.

Vol. 21 Katarzyna Dziubalska-Kolaczyk & Joanna Przedlacka (eds)
English Pronunciation Models: A Changing Scene.
2nd edition. 476 pages. 2005, 2008. ISBN 978-3-03911-682-9.

Vol. 22 Christián Abello-Contesse, Rubén Chacón-Beltrán, M. Dolores López-Jiménez & M. Mar Torreblanca-López (eds)
Age in L2 Acquisition and Teaching.
214 pages. 2006. ISBN 3-03910-668-6 · US-ISBN 0-8204-7174-7

Vol. 23 Vijay K. Bhatia, Maurizio Gotti, Jan Engberg & Dorothee Heller (eds)
Vagueness in Normative Texts.
474 pages. 2005. ISBN 3-03910-653-8 · US-ISBN 0-8204-7169-0

Vol. 24 Paul Gillaerts & Maurizio Gotti (eds)
Genre Variation in Business Letters. 2nd printing.
407 pages. 2008. ISBN 978-3-03911-681-2.

Vol. 25 Ana María Hornero, María José Luzón & Silvia Murillo (eds)
Corpus Linguistics. Applications for the Study of English.
2nd printing. 526 pages. 2006, 2008. ISBN 978-3-03911-726-0

Vol. 26 J. Lachlan Mackenzie & María de los Ángeles Gómez-González (eds)
Studies in Functional Discourse Grammar.
259 pages. 2005. ISBN 3-03910-696-1 · US-ISBN 0-8204-7558-0

Vol. 27 Debbie G. E. Ho
Classroom Talk. Exploring the Sociocultural Structure of Formal ESL Learning.
2nd edition. 254 pages. 2006, 2007. ISBN 978-3-03911-434-4

Vol. 28 Javier Pérez-Guerra, Dolores González-Álvarez, Jorge L. Bueno-Alonso & Esperanza Rama-Martínez (eds)
'Of Varying Language and Opposing Creed': New Insights into Late Modern English.
455 pages. 2007. ISBN 978-3-03910-788-9

Vol. 29 Francesca Bargiela-Chiappini & Maurizio Gotti (eds)
Asian Business Discourse(s).
350 pages. 2005. ISBN 3-03910-804-2 · US-ISBN 0-8204-7574-2

Vol. 30 Nicholas Brownlees (ed.)
News Discourse in Early Modern Britain. Selected Papers of CHINED 2004.
300 pages. 2006. ISBN 3-03910-805-0 · US-ISBN 0-8204-8025-8

Vol. 31 Roberta Facchinetti & Matti Rissanen (eds)
Corpus-based Studies of Diachronic English.
300 pages. 2006. ISBN 3-03910-851-4 · US-ISBN 0-8204-8040-1

Vol. 32 Marina Dossena & Susan M. Fitzmaurice (eds)
Business and Official Correspondence: Historical Investigations.
209 pages. 2006. ISBN 3-03910-880-8 · US-ISBN 0-8204-8352-4

Vol. 33 Giuliana Garzone & Srikant Sarangi (eds)
Discourse, Ideology and Specialized Communication.
494 pages. 2007. ISBN 978-3-03910-888-6

Vol. 34 Giuliana Garzone & Cornelia Ilie (eds)
The Use of English in Institutional and Business Settings.
An Intercultural Perspective.
372 pages. 2007. ISBN 978-3-03910-889-3

Vol. 35 Vijay K. Bhatia & Maurizio Gotti (eds)
Explorations in Specialized Genres.
316 pages. 2006. ISBN 3-03910-995-2 · US-ISBN 0-8204-8372-9

Vol. 36 Heribert Picht (ed.)
Modern Approaches to Terminological Theories and Applications.
432 pages. 2006. ISBN 3-03911-156-6 · US-ISBN 0-8204-8380-X

Vol. 37 Anne Wagner & Sophie Cacciaguidi-Fahy (eds)
Legal Language and the Search for Clarity / Le langage juridique et la quête de clarté.
Practice and Tools / Pratiques et instruments.
487 pages. 2006. ISBN 3-03911-169-8 · US-ISBN 0-8204-8388-5

Vol. 38 Juan Carlos Palmer-Silveira, Miguel F. Ruiz-Garrido &
Inmaculada Fortanet-Gómez (eds)
Intercultural and International Business Communication:
Theory, Research and Teaching.
2nd edition. 343 pages. 2006, 2008. ISBN 978-3-03911-680-5

Vol. 39 Christiane Dalton-Puffer, Dieter Kastovsky, Nikolaus Ritt &
Herbert Schendl (eds)
Syntax, Style and Grammatical Norms: English from 1500–2000.
250 pages. 2006. ISBN 3-03911-181-7 · US-ISBN 0-8204-8394-X

Vol. 40 Marina Dossena & Irma Taavitsainen (eds)
Diachronic Perspectives on Domain-Specific English.
280 pages. 2006. ISBN 3-03910-176-0 · US-ISBN 0-8204-8391-5

Vol. 41 John Flowerdew & Maurizio Gotti (eds)
Studies in Specialized Discourse.
293 pages. 2006. ISBN 3-03911-178-7

Vol. 42 Ken Hyland & Marina Bondi (eds)
Academic Discourse Across Disciplines.
320 pages. 2006. ISBN 3-03911-183-3 · US-ISBN 0-8204-8396-6

Vol. 43 Paul Gillaerts & Philip Shaw (eds)
The Map and the Landscape: Norms and Practices in Genre.
256 pages. 2006. ISBN 3-03911-182-5 · US-ISBN 0-8204-8395-4

Vol. 44 Maurizio Gotti & Davide Giannoni (eds)
New Trends in Specialized Discourse Analysis.
301 pages. 2006. ISBN 3-03911-184-1 · US-ISBN 0-8204-8381-8

Vol. 45 Maurizio Gotti & Françoise Salager-Meyer (eds)
Advances in Medical Discourse Analysis: Oral and Written Contexts.
492 pages. 2006. ISBN 3-03911-185-X · US-ISBN 0-8204-8382-6

Vol. 46 Maurizio Gotti & Susan Šarcević (eds)
Insights into Specialized Translation.
396 pages. 2006. ISBN 3-03911-186-8 · US-ISBN 0-8204-8383-4

Vol. 47 Khurshid Ahmad & Margaret Rogers (eds)
Evidence-based LSP: Translation, Text and Terminology.
584 pages. 2007. ISBN 978-3-03911-187-9

Vol. 48 Hao Sun & Dániel Z. Kádár (eds)
It's the Dragon's Turn: Chinese Institutional Discourses.
262 pages. 2008. ISBN 978-3-03911-175-6

Vol. 49 Cristina Suárez-Gómez
Relativization in Early English (950-1250): the Position of Relative Clauses.
149 pages. 2006. ISBN 3-03911-203-1 · US-ISBN 0-8204-8904-2

Vol. 50 Maria Vittoria Calvi & Luisa Chierichetti (eds)
Nuevas tendencias en el discurso de especialidad.
319 pages. 2006. ISBN 978-3-03911-261-6

Vol. 51 Mari Carmen Campoy & María José Luzón (eds)
Spoken Corpora in Applied Linguistics.
274 pages. 2008. ISBN 978-3-03911-275-3

Vol. 52 Konrad Ehlich & Dorothee Heller (Hrsg.)
Die Wissenschaft und ihre Sprachen.
323 pages. 2006. ISBN 978-3-03911-272-2

Vol. 53 Jingyu Zhang
The Semantic Salience Hierarchy Model: The L2 Acquisition of Psych Predicates
273 pages. 2007. ISBN 978-3-03911-300-2

Vol. 54 Norman Fairclough, Giuseppina Cortese & Patrizia Ardizzone (eds)
Discourse and Contemporary Social Change.
555 pages. 2007. ISBN 978-3-03911-276-0

Vol. 55 Jan Engberg, Marianne Grove Ditlevsen, Peter Kastberg & Martin Stegu (eds)
New Directions in LSP Teaching.
331 pages. 2007. ISBN 978-3-03911-433-7

Vol. 56 Dorothee Heller & Konrad Ehlich (Hrsg.)
Studien zur Rechtskommunikation.
322 pages. 2007. ISBN 978-3-03911-436-8

Vol. 57 Forthcoming.

Vol. 58 Carmen Frehner
Email – SMS – MMS
294 pages. 2008. ISBN 978-3-03911-451-1

Vol. 59 Isabel Balteiro
The Directionality of Conversion in English: A Dia-Synchronic Study.
276 pages. 2007. ISBN 978-3-03911-241-8

Vol. 60 Maria Milagros Del Saz Rubio
English Discourse Markers of Reformulation.
237 pages. 2007. ISBN 978-3-03911-196-1

Vol. 61 Sally Burgess & Pedro Martín-Martín (eds)
English as an Additional Language in Research Publication and Communication.
259 pages. 2008. ISBN 978-3-03911-462-7

Vol. 62 Sandrine Onillon
Pratiques et représentations de l'écrit.
458 pages. 2008. ISBN 978-3-03911-464-1

Vol. 63 Hugo Bowles & Paul Seedhouse (eds)
Conversation Analysis and Language for Specific Purposes.
2nd edition. 337 pages. 2007, 2009. ISBN 978-3-0343-0045-2

Vol. 64 Vijay K. Bhatia, Christopher N. Candlin & Paola Evangelisti Allori (eds)
Language, Culture and the Law.
The Formulation of Legal Concepts across Systems and Cultures.
342 pages. 2008. ISBN 978-3-03911-470-2

Vol. 65 Jonathan Culpeper & Dániel Z. Kádár (eds)
Historical (Im)politeness.
300 pages. 2010. ISBN 978-3-03911-496-2

Vol. 66 Linda Lombardo (ed.)
Using Corpora to Learn about Language and Discourse.
237 pages. 2009. ISBN 978-3-03911-522-8

Vol. 67 Natsumi Wakamoto
Extroversion/Introversion in Foreign Language Learning.
Interactions with Learner Strategy Use.
159 pages. 2009. ISBN 978-3-03911-596-9

Vol. 68 Eva Alcón-Soler (ed.)
Learning How to Request in an Instructed Language Learning Context.
260 pages. 2008. ISBN 978-3-03911-601-0

Vol. 69 Domenico Pezzini
The Translation of Religious Texts in the Middle Ages.
428 pages. 2008. ISBN 978-3-03911-600-3

Vol. 70 Tomoko Tode
Effects of Frequency in Classroom Second Language Learning.
Quasi-experiment and stimulated-recall analysis.
195 pages. 2008. ISBN 978-3-03911-602-7

Vol. 71 Egor Tsedryk
Fusion symétrique et alternances ditransitives.
211 pages. 2009. ISBN 978-3-03911-609-6

Vol. 72 Cynthia J. Kellett Bidoli & Elana Ochse (eds)
English in International Deaf Communication.
444 pages. 2008. ISBN 978-3-03911-610-2

Vol. 73 Joan C. Beal, Carmela Nocera & Massimo Sturiale (eds)
Perspectives on Prescriptivism.
269 pages. 2008. ISBN 978-3-03911-632-4

Vol. 74 Carol Taylor Torsello, Katherine Ackerley & Erik Castello (eds)
Corpora for University Language Teachers.
308 pages. 2008. ISBN 978-3-03911-639-3

Vol. 75 María Luisa Pérez Cañado (ed.)
English Language Teaching in the European Credit Transfer System:
Facing the Challenge.
251 pages. 2009. ISBN 978-3-03911-654-6

Vol. 76 Marina Dossena & Ingrid Tieken-Boon van Ostade (eds)
Studies in Late Modern English Correspondence. Methodology and Data.
291 pages. 2008. ISBN 978-3-03911-658-4

Vol. 77 Ingrid Tieken-Boon van Ostade & Wim van der Wurff (eds)
Current Issues in Late Modern English.
436 pages. 2009. ISBN 978-3-03911-660-7

Vol. 78 Marta Navarro Coy (ed.)
Practical Approaches to Foreign Language Teaching and Learning.
297 pages. 2009. ISBN 978-3-03911-661-4

Vol. 79 Qing Ma
Second Language Vocabulary Acquisition.
333 pages. 2009. ISBN 978-3-03911-666-9

Vol. 80 Martin Solly, Michelangelo Conoscenti & Sandra Campagna (eds)
Verbal/Visual Narrative Texts in Higher Education.
384 pages. 2008. ISBN 978-3-03911-672-0

Vol. 81 Meiko Matsumoto
From Simple Verbs to Periphrastic Expressions:
The Historical Development of Composite Predicates, Phrasal Verbs,
and Related Constructions in English.
235 pages. 2008. ISBN 978-3-03911-675-1

Vol. 82 Melinda Dooly
Doing Diversity. Teachers' Construction of Their Classroom Reality.
180 pages. 2009. ISBN 978-3-03911-687-4

Vol. 83 Victoria Guillén-Nieto, Carmen Marimón-Llorca & Chelo Vargas-Sierra (eds)
Intercultural Business Communication and
Simulation and Gaming Methodology.
392 pages. 2009. ISBN 978-3-03911-688-1

Vol. 84 Maria Grazia Guido
English as a Lingua Franca in Cross-cultural Immigration Domains.
285 pages. 2008. ISBN 978-3-03911-689-8

Vol. 85 Erik Castello
Text Complexity and Reading Comprehension Tests.
352 pages. 2008. ISBN 978-3-03911-717-8

Vol. 86 Maria-Lluisa Gea-Valor, Isabel García-Izquierdo & Maria-José Esteve (eds)
Linguistic and Translation Studies in Scientific Communication.
317 pages. 2010. ISBN 978-3-0343-0069-8

Vol. 87 Carmen Navarro, Rosa Mª Rodríguez Abella, Francesca Dalle Pezze
& Renzo Miotti (eds)
La comunicación especializada.
355 pages. 2008. ISBN 978-3-03911-733-8

Vol. 88 Kiriko Sato
The Development from Case-Forms to Prepositional Constructions
in Old English Prose.
231 pages. 2009. ISBN 978-3-03911-763-5

Vol. 89 Dorothee Heller (Hrsg.)
Formulierungsmuster in deutscher und italienischer Fachkommunikation. Intra- und interlinguale Perspektiven.
315 pages. 2008. ISBN 978-3-03911-778-9

Vol. 90 Henning Bergenholtz, Sandro Nielsen & Sven Tarp (eds)
Lexicography at a Crossroads. Dictionaries and Encyclopedias Today, Lexicographical Tools Tomorrow.
372 pages. 2009. ISBN 978-3-03911-799-4

Vol. 91 Manouchehr Moshtagh Khorasani
The Development of Controversies: From the Early Modern Period to Online Discussion Forums.
317 pages. 2009. ISBN 978-3-3911-711-6

Vol. 92 María Luisa Carrió-Pastor (ed.)
Content and Language Integrated Learning: Cultural Diversity.
178 pages. 2009. ISBN 978-3-3911-818-2

Vol. 93 Roger Berry
Terminology in English Language Teaching: Nature and Use.
262 pages. 2010. ISBN 978-3-0343-0013-1

Vol. 94 Roberto Cagliero & Jennifer Jenkins (eds)
Discourses, Communities, and Global Englishes
240 pages. 2010. ISBN 978-3-0343-0012-4

Vol. 95 Facchinetti Roberta, Crystal David, Seidlhofer Barbara (eds)
From International to Local English – And Back Again.
268 pages. 2010. ISBN 978-3-0343-0011-7

Vol. 96 Cesare Gagliardi & Alan Maley (eds)
EIL, ELF, Global English: Teaching and Learning Issues
376 pages. 2010. ISBN 978-3-0343-0010-0

Vol. 97 Sylvie Hancil (ed.)
The Role of Prosody in Affective Speech.
403 pages. 2009. ISBN 978-3-03911-696-6

Vol. 98 Marina Dossena & Roger Lass (eds)
Studies in English and European Historical Dialectology.
257 pages. 2009. ISBN 978-3-0343-0024-7

Vol. 99 Christine Béal
Les interactions quotidiennes en français et en anglais. De l'approche comparative à l'analyse des situations interculturelles.
424 pages. 2010. ISBN 978-3-0343-0027-8

Vol. 100 Maurizio Gotti (ed.)
Commonality and Individuality in Academic Discourse.
398 pages. 2009. ISBN 978-3-0343-0023-0

Vol. 101 Javier E. Díaz Vera & Rosario Caballero (eds)
Textual Healing: Studies in Medieval English Medical, Scientific and Technical Texts.
213 pages. 2009. ISBN 978-3-03911-822-9

Vol. 102 Nuria Edo Marzá
The Specialised Lexicographical Approach: A Step further in Dictionary-making.
316 pages. 2009. ISBN 978-3-0343-0043-8

Vol. 103 Carlos Prado-Alonso, Lidia Gómez-García, Iria Pastor-Gómez & David Tizón-Couto (eds)
New Trends and Methodologies in Applied English Language Research. Diachronic, Diatopic and Contrastive Studies.
348 pages. 2009. ISBN 978-3-0343-0046-9

Vol. 104 Françoise Salager-Meyer & Beverly A. Lewin
Crossed Words: Criticism in Scholarly Writing?
371 pages. 2011. ISBN 978-3-0343-0049-0.

Vol. 105 Javier Ruano-García
Early Modern Northern English Lexis: A Literary Corpus-Based Study.
611 pages. 2010. ISBN 978-3-0343-0058-2

Vol. 106 Rafael Monroy-Casas
Systems for the Phonetic Transcription of English: Theory and Texts.
280 pages. 2011. ISBN 978-3-0343-0059-9

Vol. 107 Nicola T. Owtram
The Pragmatics of Academic Writing.
A Relevance Approach to the Analysis of Research Article Introductions.
311 pages. 2009. ISBN 978-3-0343-0060-5

Vol. 108 Yolanda Ruiz de Zarobe, Juan Manuel Sierra & Francisco Gallardo del Puerto (eds)
Content and Foreign Language Integrated Learning.
Contributions to Multilingualism in European Contexts
343 pages. 2011. ISBN 978-3-0343-0074-2

Vol. 109 Ángeles Linde López & Rosalía Crespo Jiménez (eds)
Professional English in the European context: The EHEA challenge.
374 pages. 2010. ISBN 978-3-0343-0088-9

Vol. 110 Rosalía Rodríguez-Vázquez
The Rhythm of Speech, Verse and Vocal Music: A New Theory.
394 pages. 2010. ISBN 978-3-0343-0309-5

Vol. 111 Anastasios Tsangalidis & Roberta Facchinetti (eds)
Studies on English Modality. In Honour of Frank Palmer.
392 pages. 2009. ISBN 978-3-0343-0310-1

Vol. 112 Forthcoming.

Vol. 113 Mihhail Lotman & Maria-Kristiina Lotman (eds)
Frontiers in Comparative Prosody. In memoriam: Mikhail Gasparov.
426 pages. 2011. ISBN 978-3-0343-0373-6

Vol. 114 Merja Kytö, John Scahill & Harumi Tanabe (eds)
Language Change and Variation from Old English to Late Modern English. A Festschrift for Minoji Akimoto
422 pages. 2010. ISBN 978-3-0343-0372-9

Vol. 115 Giuliana Garzone & Paola Catenaccio (eds)
Identities across Media and Modes: Discursive Perspectives.
379 pages. 2009. ISBN 978-3-0343-0386-6

Vol. 116 Elena Landone
Los marcadores del discurso y cortesía verbal en español.
390 pages. 2010. ISBN 978-3-0343-0413-9

Vol. 117 Maurizio Gotti & Christopher Williams (eds)
Legal Discourse across Languages and Cultures.
339 pages. 2010. ISBN 978-3-0343-0425-2

Vol. 118 David Hirsh
Academic Vocabulary in Context.
217 pages. 2010. ISBN 978-3-0343-0426-9

Vol. 119 Yvonne Dröschel
Lingua Franca English. The Role of Simplification and Transfer.
358 pages. 2011. ISBN 978-3-0343-0432-0

Vol. 120 Tengku Sepora Tengku Mahadi, Helia Vaezian & Mahmoud Akbari
Corpora in Translation. A Practical Guide.
135 pages. 2010. ISBN 978-3-0343-0434-4

Vol. 121 Davide Simone Giannoni & Celina Frade (eds)
Researching Language and the Law. Textual Features and Translation Issues.
278 pages. 2010. ISBN 978-3-0343-0443-6

Vol. 122 Daniel Madrid & Stephen Hughes (eds)
Studies in Bilingual Education.
472 pages. 2011. ISBN 978-3-0343-0474-0

Vol. 123 Vijay K. Bhatia, Christopher N. Candlin & Maurizio Gotti (eds)
The Discourses of Dispute Resolution.
290 pages. 2010. ISBN 978-3-0343-0476-4

Vol. 124 Davide Simone Giannoni
Mapping Academic Values in the Disciplines. A Corpus-Based Approach.
288 pages. 2010. ISBN 978-3-0343-0488-7

Vol. 125 Giuliana Garzone & James Archibald (eds)
Discourse, Identities and Roles in Specialized Communication.
419 pages. 2010. ISBN 978-3-0343-0494-8

Vol. 126 Iria Pastor-Gómez
The Status and Development of N+N Sequences in
Contemporary English Noun Phrases.
216 pages. 2011. ISBN 978-3-0343-0534-1

Vol. 127 Carlos Prado-Alonso
Full-verb Inversion in Written and Spoken English.
261 pages. 2011. ISBN 978-3-0343-0535-8

Vol. 128 Tony Harris & María Moreno Jaén (eds)
Corpus Linguistics in Language Teaching.
214 pages. 2010. ISBN 978-3-0343-0524-2

Vol. 129 Tetsuji Oda & Hiroyuki Eto (eds)
Multiple Perspectives on English Philology and History of Linguistics.
A Festschrift for Shoichi Watanabe on his 80th Birthday.
378 pages. 2010. ISBN 978-3-0343-0480-1

Vol. 130 Luisa Chierichetti & Giovanni Garofalo (eds)
Lengua y Derecho: líneas de investigación interdisciplinaria.
283 pages. 2010. 978-3-0343-0463-4

Vol. 131 Paola Evangelisti Allori & Giuliana Garzone (eds)
Discourse, Identities and Genres in Corporate Communication.
Sponsorship, Advertising and Organizational Communication.
324 pages. 2011. 978-3-0343-0591-4

Vol. 132 Leyre Ruiz de Zarobe & Yolanda Ruiz de Zarobe (eds)
Speech Acts and Politeness across Languages and Cultures.
402 pages. 2012. 978-3-0343-0611-9

Vol. 133 Thomas Christiansen
Cohesion: A Discourse Perspective.
387 pages. 2011. 978-3-0343-0619-5

Vol. 134 Giuliana Garzone & Maurizio Gotti
Discourse, Communication and the Enterprise. Genres and Trends.
451 pages. 2011. ISBN 978-3-0343-0620-1

Vol. 135 Zsuzsa Hoffmann
Ways of the World's Words.
Language Contact in the Age of Globalization.
334 pages 2011. ISBN 978-3-0343-0673-7

Vol. 136 Cecilia Varcasia (ed.)
Becoming Multilingual.
Language Learning and Language Policy between Attitudes and Identities.
213 pages. 2011. ISBN 978-3-0343-0687-5

Vol. 137 Susy Macqueen
The Emergence of Patterns in Second Language Writing.
A Sociocognitive Exploration of Lexical Trails.
325 pages. 2012. ISBN 978-3-0343-1010-9

Vol. 138 Maria Vittoria Calvi & Giovanna Mapelli (eds)
La lengua del turismo. Géneros discursivos y terminología.
365 pages. 2011. ISBN 978-3-0343-1011-6

Vol. 139 Ken Lau
Learning to Become a Professional in a Textually-Mediated World.
A Text-Oriented Study of Placement Practices.
261 pages. 2012. ISBN 978-3-0343-1016-1

Vol. 140 Sandra Campagna, Giuliana Garzone, Cornelia Ilie & Elizabeth Rowley-Jolivet (eds)
Evolving Genres in Web-mediated Communication.
337 pages. 2012. ISBN 978-3-0343-1013-0

Vol. 141 Edith Esch & Martin Solly (eds)
The Sociolinguistics of Language Education in International Contexts.
263 pages. 2012. ISBN 978-3-0343-1009-3

Vol. 142 Forthcoming.

Vol. 143 David Tizón-Couto
Left Dislocation in English. A Functional-Discoursal Approach.
416 pages. 2012. ISBN 978-3-0343-1037-6

Vol. 144 Margrethe Petersen & Jan Engberg (eds)
Current Trends in LSP Research. Aims and Methods.
323 pages. 2011. ISBN 978-3-0343-1054-3

Vol. 145 David Tizón-Couto, Beatriz Tizón-Couto, Iria Pastor-Gómez & Paula Rodríguez-Puente (eds)
New Trends and Methodologies in Applied English Language Research II.
Studies in Language Variation, Meaning and Learning.
283 pages. 2012. ISBN 978-3-0343-1061-1

Vol. 146 Rita Salvi & Hiromasa Tanaka (eds)
Intercultural Interactions in Business and Management.
306 pages. 2011. ISBN 978-3-0343-1039-0

Vol. 147 Francesco Straniero Sergio & Caterina Falbo (eds)
Breaking Ground in Corpus-based Interpreting Studies.
254 pages. 2012. ISBN 978-3-0343-1071-0

Vol. 148 Forthcoming.

Vol. 149 Vijay K. Bhatia & Paola Evangelisti Allori (eds)
Discourse and Identity in the Professions. Legal, Corporate and Institutional Citizenship.
352 pages. 2011. ISBN 978-3-0343-1079-6

Vol. 150 Maurizio Gotti (ed.)
Academic Identity Traits. A Corpus-Based Investigation.
363 pages. 2012. ISBN 978-3-0343-1141-0

Vol. 151 Priscilla Heynderickx, Sylvain Dieltjens, Geert Jacobs, Paul Gillaerts &
Elizabeth de Groot (eds)
The Language Factor in International Business.
New Perspectives on Research, Teaching and Practice.
320 pages. 2012. ISBN 978-3-0343-1090-1

Vol. 152 Paul Gillaerts, Elizabeth de Groot, Sylvain Dieltjens, Priscilla Heynderickx &
Geert Jacobs (eds)
Researching Discourse in Business Genres. Cases and Corpora.
215 pages. 2012. ISBN 978-3-0343-1092-5

Vol. 153 Yongyan Zheng
Dynamic Vocabulary Development in a Foreign Language.
262 pages. 2012. ISBN 978-3-0343-1106-9

Vol. 154 Forthcoming.

Vol. 155 David Hirsh (ed.)
Current Perspectives in Second Language Vocabulary Research.
180 pages. 2012. ISBN 978-3-0343-1108-3

Vol. 156 Seiji Shinkawa
Unhistorical Gender Assignment in Laȝamon's *Brut.* A Case Study of a Late Stage
in the Development of Grammatical Gender toward its Ultimate Loss.
186 pages. 2012. ISBN 978-3-0343-1124-3

Vol. 157 Forthcoming.

Vol. 158 Bárbara Eizaga Rebollar (ed.)
Studies in Linguistics and Cognition.
301 pages. 2012. ISBN 978-3-0343-1138-0

Vol. 159 Giuliana Garzone, Paola Catenaccio, Chiara Degano (eds)
Genre Change in the Contemporary World. Short-term Diachronic Perspectives.
329 pages. 2012. ISBN 978-3-0343-1214-1

Vol. 160 Carol Berkenkotter, Vijay K. Bhatia & Maurizio Gotti (eds)
Insights into Academic Genres.
468 pages. 2012. ISBN 978-3-0343-1211-0

Vol. 161 Forthcoming.

Vol. 162 Patrizia Anesa
Jury Trials and the Popularization of Legal Language. A Discourse Analytical Approach.
247 pages. 2012. ISBN 978-3-0343-1231-8

www.ingramcontent.com/pod-product-compliance
Ingram Content Group UK Ltd.
Pitfield, Milton Keynes, MK11 3LW, UK
UKHW021853100926
8556IPUK00020B/1199

9 783034 310130